Active Reading Note-Taking G

Glencoe

World Geography

Student Workbook

Douglas Fisher, Ph.D.
San Diego State University

 Glencoe

New York, New York   Columbus, Ohio   Chicago, Illinois   Peoria, Illinois   Woodland Hills, California

About the Author

Douglas Fisher, Ph.D., is a Professor in the Department of Teacher Education at San Diego State University. He is the recipient of an International Reading Association Celebrate Literacy Award as well as a Christa McAuliffe award for excellence in teacher education. He has published numerous articles on reading and literacy, differentiated instruction, and curriculum design as well as books, such as *Improving Adolescent Literacy: Strategies at Work* and *Responsive Curriculum Design in Secondary Schools: Meeting the Diverse Needs of Students.* He has taught a variety of courses in SDSU's teacher-credentialing program as well as graduate-level courses on English language development and literacy. He has also taught classes in English, writing, and literacy development to secondary school students.

The McGraw-Hill Companies

Send all inquiries to:
Glencoe/McGraw-Hill
8787 Orion Place
Columbus, Ohio 43240-4027

ISBN 0-07-867989-3

Printed in the United States of America

4 5 6 7 8 9 10   045   09 08 07 06

Table of Contents

Letter to the Student vii

Chapter 1: **How Geographer's Look at the World** 1

Section 1: Exploring Geography 1

Section 2: The Geographer's Craft 7

Chapter 2: **The Earth** 12

Section 1: Planet Earth 12

Section 2: Forces of Change 17

Section 3: Earth's Water 22

Chapter 3: **Climates of the Earth** 27

Section 1: Earth-Sun Relationships 27

Section 2: Factors Affecting Climate 33

Section 3: World Climate Patterns 38

Chapter 4: **The Human World** 42

Section 1: World Populations 42

Section 2: Global Cultures 47

Section 3: Political and Economic Systems 52

Section 4: Resources, Trade, and the Environment 57

Chaper 5: **The Physical Geography of the United States and Canada** 63

Section 1: The Land 63

Section 2: Climate and Vegetation...68

Chapter 6: **The Cultural Geography of the United States and Canada** 74

Section 1: Population Patterns 74

Section 2: History and Government 79

Section 3: Cultures and Lifestyles 84

Chapter 7: **The United States and Canada Today** 89

Section 1: Living in the United States and Canada 89

Section 2: People and Their Environment...................................95

Chapter 8: **The Physical Geography of Latin America** 100

Section 1: The Land 100

Section 2: Climate and Vegetation...106

Table of Contents

Chapter 9: The Cultural Geography of Latin America 111
- **Section 1:** Population Patterns 111
- **Section 2:** History and Government 117
- **Section 3:** Cultures and Lifestyles 122

Chapter 10: Latin America Today 127
- **Section 1:** Living in Latin America 127
- **Section 2:** People and Their Environment.......133

Chapter 11: The Physical Geography of Europe 138
- **Section 1:** The Land 138
- **Section 2:** Climate and Vegetation...............143

Chapter 12: The Cultural Geography of Europe 148
- **Section 1:** Population Patterns 148
- **Section 2:** History and Government................153
- **Section 3:** Cultures and Lifestyles 159

Chapter 13: Europe Today 164
- **Section 1:** Living in Europe 164
- **Section 2:** People and Their Environment.......169

Chapter 14: The Physical Geography of Russia 174
- **Section 1:** The Land 174
- **Section 2:** Climate and Vegetation...............179

Chapter 15: The Cultural Geography of Russia 184
- **Section 1:** Population Patterns 184
- **Section 2:** History and Government 188
- **Section 3:** Cultures and Lifestyles 194

Chapter 16: Russia Today 199
- **Section 1:** Living in Russia...........................199
- **Section 2:** People and Their Environment.......203

Chapter 17: The Physical Geography of North America, Southwest Asia, and Central Asia 208
- **Section 1:** The Land 208
- **Section 2:** Climate and Vegetation................212

Table of Contents

**Chapter 18: The Cultural Geography of North America,
Southwest Asia, and Central Asia**..217
  Section 1: Population Patterns 217
  Section 2: History and Government 221
  Section 3: Cultures and Lifestyles 226

Chapter 19: North America, Southwest Asia, and Central Asia Today 231
  Section 1: Living in North America, Southwest Asia, and Central Asia 231
  Section 2: People and Their Environment...236

Chapter 20: The Physical Geography of Africa South of the Sahara 240
  Section 1: The Land 240
  Section 2: Climate and Vegetation...245

Chapter 21: The Cultural Geography of Africa South of the Sahara...............249
  Section 1: Population Patterns 249
  Section 2: History and Government 253
  Section 3: Cultures and Lifestyles 258

Chapter 22: Africa South of the Sahara Today....................................264
  Section 1: Living in Africa South of the Sahara....................................264
  Section 2: People and Their Environment...269

Chapter 23: The Physical Geography of South Asia 273
  Section 1: The Land 273
  Section 2: Climate and Vegetation...278

Chapter 24: The Cultural Geography of South Asia 282
  Section 1: Population Patterns 282
  Section 2: History and Government 286
  Section 3: Cultures and Lifestyles 291

Chapter 25: South Asia Today 296
  Section 1: Living in South Asia 296
  Section 2: People and Their Environment...301

Chapter 26: The Physical Geography of East Asia 306
  Section 1: The Land 306
  Section 2: Climate and Vegetation...311

Table of Contents

Chapter 27: The Cultural Geography of East Asia 316
  Section 1: Population Patterns 316
  Section 2: History and Government 321
  Section 3: Cultures and Lifestyles 326

Chapter 28: East Asia Today 331
  Section 1: Living in East Asia 331
  Section 2: People and Their Environment 336

Chapter 29: The Physical Geography of Southeast Asia 341
  Section 1: The Land 341
  Section 2: Climate and Vegetation 345

Chapter 30: The Cultural Geography of Southeast Asia 349
  Section 1: Population Patterns 349
  Section 2: History and Government 354
  Section 3: Cultures and Lifestyles 359

Chapter 31: Southeast Asia Today 363
  Section 1: Living in Southeast Asia 363
  Section 2: People and Their Environment 369

Chapter 32: The Physical Geography of Australia, Oceania, and Antarctica 374
  Section 1: The Land 374
  Section 2: Climate and Vegetation 379

Chapter 33: The Cultural Geography of Australia, Oceania, and Antarctica 384
  Section 1: Population Patterns 384
  Section 2: History and Government 389
  Section 3: Cultures and Lifestyles 394

Chapter 34: Australia, Oceania, and Antarctica Today 398
  Section 1: Living in Australia, Oceania, and Antarctica 398
  Section 2: People and Their Environment 403

Dear Social Studies Student,

Can you believe it? The start of another school year is upon you. How exciting to be learning about different cultures, historical events, and unique places in your social studies class! I believe that this Active Reading Note-Taking Guide *will help you as you learn about your community, nation, and world.*

Note-Taking and Student Success

Did you know that the ability to take notes helps you become a better student? Research suggests that good notes help you become more successful on tests because the act of taking notes helps you remember and understand content. This *Active Reading Note-Taking Guide* is a tool that you can use to achieve this goal. I'd like to share some of the features of this *Active Reading Note-Taking Guide* with you before you begin your studies.

The Cornell Note-Taking System

First, you will notice that the pages in the *Active Reading Note-Taking Guide* are arranged in two columns, which will help you organize your thinking. This two-column design is based on the **Cornell Note-Taking System**, developed at Cornell University. The column on the left side of the page highlights the main ideas and vocabulary of the lesson. This column will help you find information and locate the references in your textbook quickly. You can also use this column to sketch drawings that further help you visually remember the lesson's information. In the column on the right side of the page, you will write detailed notes about the main ideas and vocabulary. The notes you take in this column will help you focus on the important information in the lesson. As you become more comfortable using the **Cornell Note-Taking System**, you will see that it is an important tool that helps you organize information.

The Importance of Graphic Organizers

Second, there are many graphic organizers in this *Active Reading Note-Taking Guide*. Graphic organizers allow you to see the lesson's important information in a visual format. In addition, graphic organizers help you understand and summarize information, as well as remember the content.

Research-Based Vocabulary Development

Third, you will notice that vocabulary is introduced and practiced throughout the *Active Reading Note-Taking Guide*. When you know the meaning of the words used to discuss information, you are able to understand that information better. Also, you are more likely to be successful in school when you have vocabulary knowledge. When researchers study successful students, they find that as students acquire vocabulary knowledge, their ability to learn improves. The *Active Reading Note-Taking Guide* focuses on learning words that are very specific to understanding the content of your textbook. It also highlights general academic words that you need to know so that you can understand any textbook. Learning new vocabulary words will help you succeed in school.

Writing Prompts and Note-Taking

Finally, there are a number of writing exercises included in this *Active Reading Note-Taking Guide*. Did you know that writing helps you to think more clearly? It's true. Writing is a useful tool that helps you know if you understand the information in your textbook. It helps you assess what you have learned.

You will see that many of the writing exercises require you to practice the skills of good readers. Good readers *make connections* between their lives and the text and *predict* what will happen next in the reading. They *question* the information and the author of the text, *clarify* information and ideas, and *visualize* what the text is saying. Good readers also *summarize* the information that is presented and *make inferences* or *draw conclusions* about the facts and ideas.

I wish you well as you begin another school year. This *Active Reading Note-Taking Guide* is designed to help you understand the information in your social studies class. The guide will be a valuable tool that will also provide you with skills you can use throughout your life.

I hope you have a successful school year.

Sincerely,

Douglas Fisher

Chapter 1, Section 1
Exploring Geography
(Pages 19–22)

Reason To Read

Setting a Purpose for Reading Think about these questions as you read:
- What are the physical and human features geographers study?
- How do geographers describe the earth's features and their patterns?
- How is geography used?

Main Idea

As you read pages 19–22 in your textbook, complete this graphic organizer by listing three types of regions.

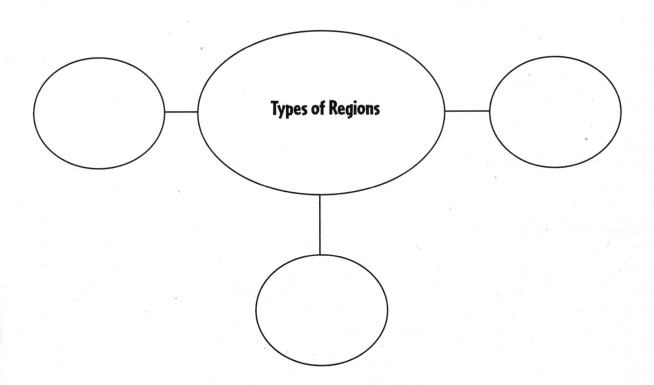

Types of Regions

The Elements of Geography *(pages 19–20)*

Summarizing

As you read, complete the following sentences. Doing so will help you summarize the section.

1. Geographers are specialists who describe the earth's _____ and _____ features

2. Geographers describe the interactions of _____ , _____ , and _____ .

3. In their work, geographers study the world in terms of six elements that include _____ , _____ , _____ , _____ , _____ , and the uses of geography.

Academic Vocabulary

Define the following academic vocabulary words.

feature ⟩ _____
_____
_____

environment ⟩ _____
_____

The World in Spatial Terms *(page 20)*

Determining the Main Idea

As you read, write the main idea of the passage. Review your statement when you have finished reading and revise as needed.

_____
_____
_____
_____
_____

Key Points

Notes

Terms To Know

Write the letter of the correct definition in Group B next to the correct term in Group A. One definition will not be used.

Group A

_____ **1.** location

_____ **2.** absolute location

_____ **3.** hemisphere

_____ **4.** grid system

_____ **5.** relative location

Group B

a. the exact spot at which a place is found on the globe

b. the pattern formed as the lines of latitude and longitude cross one another

c. the study of the earth

d. a specific place on earth

e. location in relation to other places

f. half of a sphere or globe

Academic Vocabulary

Use the following terms from this lesson in a sentence that shows you understand the term's meaning.

globe >
_____
_____

network >
_____
_____

Places and Regions *(page 21)*

Visualizing

Visualize the information described in this section to help you understand and remember what you have read. First, read the section. Next, ask yourself, What would this look like? *Finally, write a description of the pictures you visualized on the lines below.*

_____
_____
_____

Terms To Know

Define or describe the following key terms.

place >
_____
_____

Key Points	Notes

> regions

> formal region

> functional region

> perceptual region

Physical Systems *(page 21)*

Questioning

As you read, write two questions about the main ideas presented in the text. After you have finished reading, write the answers to these questions.

1. _____

   _____

2. _____

   _____

Terms To Know

Define or describe the following key term.

> ecosystem

Academic Vocabulary

Circle the letter of the word that has the closest meaning to the underlined word.

Geographer's <u>analyze</u> how physical features interact with plant and animal life.

a. accept   **b.** examine   **c.** change

Human Systems *(page 22)*

Summarizing

As you read, complete the following sentences. Doing so will help you summarize the section.

1. Geographers study how people _____ the earth,

   _____ societies, and _____ permanent features.

2. Geographers look at how people _____ or _____ to change aspects of the earth to meet their needs.

Terms To Know

Define or describe the following key term.

movement > _____

_____

Environment and Society *(page 22)*

Skimming

Read the title and quickly look over the section to get a general idea of the section's content. Then write a sentence or two explaining what the section is about.

_____

_____

_____

Terms To Know

Define or describe the following key term.

human-environment interaction > _____

_____

Academic Vocabulary

Use the following academic vocabulary terms from this lesson in a sentence.

physical > _____

features >

The Uses of Geography _(page 22)_

 _As you read, identify three details about the uses of geography. When you have finished reading, write a general statement using these details._

Details >

_____

_____

_____

_____

General Statement >

_____

_____

Section Wrap-up _Now that you have read the section, write the answers to the questions that were included in_ **Setting a Purpose for Reading** _at the beginning of the lesson._

What are the physical and human features geographers study?

_____

_____

How do geographers describe the earth's features and their patterns?

_____

_____

How is geography used?

_____

_____

Chapter 1, Section 2
The Geographer's Craft

(Pages 23-27)

Reason To Read

Setting a Purpose for Reading Think about these questions as you read:

- What are the major branches of geography and the topics each branch studies?
- What research methods do geographers use at work?
- How is geography related to the other subject areas?
- What kinds of geographic careers are available today?

Main Idea

As you read pages 23-27 in your textbook, complete this graphic organizer by listing the specialized research methods geographers use.

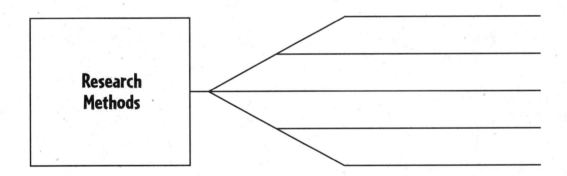

Branches of Geography (pages 23–24)

Analyzing

As you read this section, think about the section's organization and main ideas. Then write a sentence explaining the organization and list the main ideas.

Organization

_____

_____

Main Ideas

1. _____

_____

2. _____

_____

3. _____

_____

Terms To Know

Fill in the blank with the correct term from the list below. One term will not be used.

physical geography

1. The branch of geography that studies features such climate, land, water, plants, and animals is called _____ .

meteorology

2. _____ , or cultural geography, is the study of the activities of people and their relationship to the cultural and physical environments.

human geography

cultural geography

3. The study of weather and weather conditions is called

_____ .

 **Key Points**

 **Notes**

Academic Vocabulary

Define the following academic vocabulary words.

> **range**

_____
_____
_____

> **topics**

_____
_____

Geographers at Work (pages 24–26)

Previewing

Preview the section to get an idea of what is ahead. First, skim the section. Then write a sentence or two explaining what you think you will be learning. After you have finished reading, revise your statements as necessary.

_____
_____
_____

Terms To Know

Define or describe the following key terms.

> **cartography**

_____
_____

> **geographic information systems (GIS)**

_____
_____

Academic Vocabulary

Define the following academic vocabulary words.

> **research**

_____
_____

> **methods**

_____
_____

Geographers and Other Disciplines *(pages 26–27)*

Outlining   *Complete this outline as you read:*

I. History and Government

    A. _____

    B. _____

II. Culture

    A. _____

    B. _____

III. Economics

    A. _____

    B. _____

Academic Vocabulary   *Fill in the blank with the correct term from the list below. One term will be used twice.*

culture

1. _____ is the study of how people make choices about ways to use scarce resources to meet their needs and wants.

2. _____ is the way of life of a group of people who share beliefs and similar customs.

economics

3. In order to understand _____ , geographers use sociology and anthropology.

Geography as a Career *(page 27)*

Questioning   *As you read, write two questions about the main ideas presented in the text. After you have finished reading, write the answers to these questions.*

1. _____

    _____

2. _____

_____

Academic Vocabulary

Circle the letter of the word that has the closest meaning to the underlined word.

1. The <u>job</u> of a physical geographer is to study Earth's features and the geographic forces shaping them.

> job

    a. work   **b.** will   **c.** base

2. A salesperson must know the <u>region</u> in which he or she is selling goods.

> region

    a. building   **b.** area   **c.** speech

Section Wrap-up

Now that you have read the section, write the answers to the questions that were included in Setting a Purpose for Reading *at the beginning of the lesson.*

What are the major branches of geography and the topics each branch studies?

_____

_____

What research methods to geographers use at work?

_____

_____

How is geography related to the other subject areas?

_____

_____

What kinds of geographic careers are available today?

_____

_____

Chapter 2, Section 1
Planet Earth

(Pages 33–36)

Reason To Read

Setting a Purpose for Reading Think about these questions as you read:
- Where is Earth located in our solar system?
- How is Earth shaped?
- What is Earth's structure?
- What types of landforms are found on Earth?

Main Idea

As you read pages 33–36 in your textbook, complete this graphic organizer by describing the four components of Earth.

Component	Description
Hydrosphere	
Lithosphere	
Atmosphere	
Biosphere	

Our Solar System (pages 33–35)

Scanning

Scan the section before you begin to read. As you glance quickly over the lines of text, look for key words or phrases that will tell you what the text will cover. Write the key words or phrases. Then use the key words and phrases to write a statement explaining the lesson content. Revise your statement when you are finished reading the section.

Key words or phrases

_____

_____

_____

_____

What the section is about

_____

_____

_____

_____

Academic Vocabulary

Define the following academic vocabulary words from this lesson.

create

_____

_____

sphere

_____

_____

Terms To Review

Use the following term that you studied earlier in a sentence that reflects the term's meaning.

physical
(Chapter 1, Section 1)

_____

_____

Getting to Know Earth (pages 35–36)

Monitoring Comprehension

As you read the lesson, write down questions you have about what you read. When you have finished reading the lesson, answer your questions.

_____

_____

_____

_____

Terms To Know

Write the letter of the correct definition in Group B next to the correct term from this lesson in Group A. One definition will not be used.

Group A	Group A
_____ **1.** hydrosphere	**a.** the part of the earth that supports life
_____ **2.** lithosphere	**b.** seven large landmasses
_____ **3.** atmosphere	**c.** a layer of gases extending thousands of miles above Earth
_____ **4.** bioshpere	**d.** the part of the continent that extends underwater
_____ **5.** continental shelf	**e.** the part of the earth made up of bodies of water
	f. the part of the earth made up of land

Places To Locate

Fill in the blank with the correct place from this lesson.

Dead Sea   Asia   North America

Mount Everest   Australia   Europe

Mariana Trench   Antarctica   South America

Africa

1. Two continents that stand alone are _____ and

_____ .

2. _____ and _____ are actually
parts of one large landmass called Eurasia.

3. The Isthmus of Panama links the continents of _____

and _____ .

4. At the Sinai Peninsula, the human-made Suez Canal separates

_____ and _____ .

5. The highest point on Earth is in South Asia at the top of

_____ .

6. The lowest dry land point is the shore of the _____
in Southwest Asia.

7. Earth's lowest known depression lies under the Pacific Ocean
southwest of Guam in the long, underwater canyon called the

_____ .

Academic Vocabulary

Define the following academic vocabulary words from this lesson.

percent _____

_____

_____

area _____

_____

Terms To Review

Use the following terms that you studied earlier in a sentence that reflects the term's meaning.

globe
(Chapter 1, Section 1) _____

_____

feature
(Chapter 1, Section 1) _____

_____

Now that you have read the section, write the answers to the questions that were included in Setting a Purpose for Reading *at the beginning of the lesson.*

Where is Earth located in our solar system?

_____

_____

How is Earth shaped?

_____

_____

What types of landforms are found on Earth?

_____

_____

What is Earth's structure?

_____

_____

Chapter 2, Section 2
Forces of Change
(Pages 37–43)

Reason To Read

Setting a Purpose for Reading Think about these questions as you read:
- How do Earth's layers contribute to the planet's physical characteristics?
- How do internal and external forces of change affect Earth's surface differently?
- What external forces affect Earth's surface?

Main Idea

As you read pages 37–43 in your textbook, complete this graphic organizer by using the major headings of the section to create an outline similar to the one below.

Forces of Change

I. Earth's Structure

    A. _____

    B. _____

II. Internal Forces of Change

    A. _____

    B. _____

    C. _____

    D. _____

III. External Forces of Change

    A. _____

    B. _____

    C. _____

    D. _____

Earth's Structure *(pages 37–39)*

Clarifying

As you read this section, write down terms or concepts you find confusing. Reread the section to clarify the confusing terms and concepts. Write an explanation of the terms and concepts.

_____

_____

_____

Terms To Know

Define the following key terms from this lesson.

magma

_____

_____

mantle

_____

_____

plate tectonics

_____

_____

continental drift

_____

_____

Academic Vocabulary

Circle the letter of the word that has the closest meaning to the underlined word from this lesson.

occur

1. Some forces that change the earth, such as wind, <u>occur</u> on the earth's surface.

   a. develop   **b.** happen   **c.** begin

elements

2. Some of the <u>elements</u> that make up the outer core of Earth include silicon, aluminum, iron, magnesium, and oxygen.

   a. models   **b.** landforms   **c.** substances

Internal Forces of Change (pages 39–41)

Responding

As you read this section, think about what grabs your attention as you read. Write down facts you find interesting or surprising in the lesson.

_____

_____

_____

_____

Terms To Know

Write the letter of the correct definition in Group B next to the correct term in Group A. One definition will not be used.

Group A

_____ **1.** subduction

_____ **2.** accretion

_____ **3.** spreading

_____ **4.** folds

_____ **5.** faults

Group B

a. cracks in the earth's crust

b. a process in which a heavier sea plate dives beneath a lighter continental plate

c. sudden, violent movements of plates along a fault line

d. a process in which pieces of the earth's crust come together slowly as the sea plate slides under the continental plate

e. bends in layers of rock

f. a process in which sea plates pull apart

Places To Locate

Explain why each of these places from this lesson is important.

San Andreas Fault

_____

_____

Ring of Fire

_____

_____

Key Points

Notes

Academic Vocabulary

Define the following academic vocabulary words from this lesson.

minor >

_____

_____

_____

energy >

_____

_____

Terms To Review

Use the following term that you studied earlier in a sentence that reflects the term's meaning.

range >
(Chapter 1, Section 2)

_____

_____

External Forces of Change (pages 42–43)

Predicting

Read the title and main headings of the lesson. Write a statement predicting what the lesson will be about and what will be included in the text. As you read, change your prediction if it does not match what you learn.

_____

_____

Terms To Know

Choose one of the following terms from this lesson to fill in each blank. One term will not be used.

weathering   erosion   loess   glaciers   moraines

1. China's Yellow River basin is thickly covered with a fertile, yellow-gray

   soil called _____ .

2. _____ are large piles of rocks and debris
   left behind when glaciers melted and receded.

3. The process that breaks down rocks on the earth's surface into

   smaller pieces is called _____ .

4. The wearing away of the earth's surface by wind, glaciers, and moving

water is called _____ .

Academic Vocabulary

Choose one of these two academic vocabulary terms from this lesson to fill in each blank.

expands

1. Water seeps into cracks and _____ when it freezes.

altering

2. Glacier movements change the landscape by _____ the directions of rivers.

Terms To Review

Use the following terms that you studied earlier in a sentence that reflects the term's meaning.

locations
(Chapter 1, Section 1)

_____

_____

Section Wrap-up

Now that you have read the section, write the answers to the questions that were included in **Setting a Purpose for Reading** *at the beginning of the lesson.*

How do Earth's layers contribute to the planet's physical characteristics?

_____

_____

How do internal and external forces of change affect Earth's surface differently?

_____

_____

What external forces affect Earth's surface?

_____

_____

Chapter 2, Section 3
Earth's Water

(Pages 46–49)

Reason To Read

Setting a Purpose for Reading Think about these questions as you read:
- How does the amount of water on Earth remain fairly constant?
- How is the water that makes up 70 percent of the Earth's surface distributed?
- Why is freshwater important to humans?

Main Idea

As you read pages 46–49 in your textbook, complete this graphic organizer by listing the processes that contribute to the water cycle.

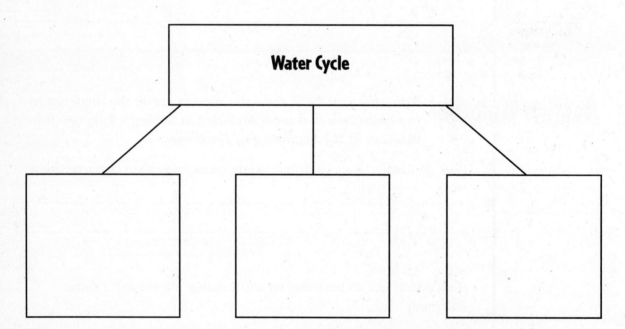

Water Cycle (pages 46–47)

Questioning

As you read, write two questions about the main ideas presented in the text. After you have finished reading, write the answers to these questions.

1. _____

_____

2. _____

_____

Terms To Know

Write the letter of the correct definition in Group B next to the correct term from this lesson in Group A. One definition will not be used.

Group A

_____ **1.** water cycle

_____ **2.** evaporation

_____ **3.** condensation

_____ **4.** precipitation

Group B

a. a process in which excess water vapor changes into liquid water

b. the regular movement of water from oceans to the air to the ground and finally back to the oceans

c. freshwater found in lakes, springs, and rivers

d. moisture in the form of rain, snow, or sleet

e. the changing of liquid water into vapor or gas

Academic Vocabulary

Circle the letter of the word that has the closest meaning to the underlined word from this lesson.

major

1. The <u>major</u> parts of the water cycle are condensation and evaporation.

   a. main   **b.** short   **c.** common

varies

2. The amount of moisture that a place gets <u>varies</u> from season to season.

   a. falls   **b.** changes   **c.** increases

Terms To Review

Use the following term that you studied earlier in a sentence that reflects the term's meaning.

hydrosphere
(Chapter 2, Section 2)

> _____
>
> _____

Bodies of Salt Water *(pages 47–48)*

Visualizing

Visualize the information described in this section to help you understand and remember what you have read. First, read the section. Next, ask yourself, **What would this look like?** *Finally, write a description of the pictures you visualized on the lines below.*

_____

_____

_____

Terms To Know

Define or describe the following key term from this lesson.

desalination

> _____
>
> _____

Places To Locate

Explain why the following places from this lesson are important.

Pacific Ocean

> _____
>
> _____

Atlantic Ocean

> _____
>
> _____

Indian Ocean

> _____
>
> _____

Arctic Ocean >

Academic Vocabulary

Define the following vocabulary words from this lesson.

identified >

sources >

Bodies of Freshwater *(pages 48–49)*

Skimming

Read the title and quickly look over the lesson to get a general idea of the lesson's content. Then write a sentence or two explaining what the lesson is about.

Terms To Know

Define or describe the following key terms from this lesson.

groundwater >

aquifer >

Academic Vocabulary

Circle the letter of the word that has the closest meaning to the underlined word from this lesson.

available

1. Since most of Earth's freshwater is frozen in glaciers and ice caps, it is not <u>available</u> for people to drink.

   a. agreeable **b.** knowledgeable **c.** obtainable

constant

2. People usually settle in places where there is a <u>constant</u> supply of water.

   a. regular **b.** fresh **c.** large

Terms To Review

Use the following terms that you studied earlier in a sentence that reflects the term's meaning.

regions
(Chapter 1, Section 1)

_____
_____
_____

place
(Chapter 1, Section 1)

_____
_____

Section Wrap-up

Now that you have read the section, write the answers to the questions that were included in Setting a Purpose for Reading *at the beginning of the lesson.*

How does the amount of water on Earth remain fairly constant?

_____
_____

How is the water that makes up 70 percent of Earth's surface distributed?

_____
_____

Why is freshwater important to humans?

_____

Chapter 3, Section 1
Earth-Sun Relationships
(Pages 55–58)

Reason To Read

Setting a Purpose for Reading Think about these questions as you read:
- How does Earth's position in relation to the sun affect temperatures on Earth?
- How does Earth's rotation cause day and night?
- What is Earth's position in relation to the sun during each season?
- How might global warming affect Earth's air, land, and water?

Main Idea

As you read pages 55–58 in your textbook, complete this graphic organizer by listing the major characteristics of the summer and winter solstices.

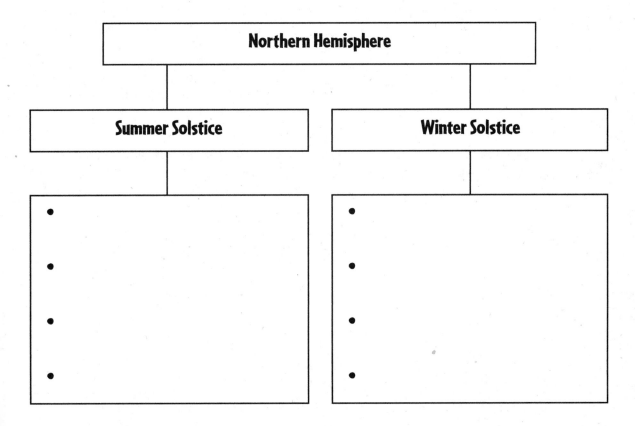

Northern Hemisphere

Summer Solstice	Winter Solstice
•	•
•	•
•	•
•	•

Climate and Weather *(pages 55–56)*

Synthesizing

As you read, think about the main ideas of the section. Ask yourself, **Do I understand more than the main ideas? Can I combine the ideas in this section to reach a new understanding?**

In which situation would you want information about the weather and in which situation would you want information about the climate? Write *weather* or *climate* on the correct blanks below.

1. _____ You are going on vacation next week to Mexico. What clothes should you pack?

2. _____ You are going to school in Europe for a year. What clothes would you pack?

3. _____ You want to travel to Australia. What time of year would be warmest?

4. _____ You want to plan a picnic at the park this week. What day will be best?

Terms To Know

Use the following terms from this lesson in a sentence that reflects the term's meaning.

weather > _____

_____

_____

climate > _____

Academic Vocabulary

Circle the letter of the word or phrase that has the closest meaning to the underlined word from this lesson.

1. Weather is a short-term <u>aspect</u> of climate.

aspect >   **a.** order   **b.** feature   **c.** advantage

Key Points

period

2. Climate is the weather pattern that an area has over a long <u>period</u> of time.

    a. stretch   **b.** start   **c.** flash

Earth's Tilt and Rotation *(page 56)*

Determining the Main Idea

As you read, write the main idea of the passage. Review your statement when you have finished reading and revise as needed.

_____

_____

Terms To Know

Define or describe the following key terms from this lesson.

axis

_____

_____

temperature

_____

_____

Earth's Revolution *(page 56)*

Visualizing

Visualize the information described in this section to help you understand and remember what you have read. First, read the section. Next, ask yourself, What would this look like? Finally, write a description of the pictures you visualized on the lines below.

_____

_____

_____

_____

Terms To Know

Define or describe the following key terms from this lesson.

revolution

_____
_____
_____

equinox

_____
_____

The Tropics of Cancer and Capricorn *(pages 56–57)*

Clarifying

As you read this section, write down terms or concepts you find confusing. Then go back and reread the section to clarify the confusing parts. Write an explanation of the terms and concepts. If you are still confused, write your questions down and ask your teacher to clarify.

_____
_____
_____

Terms To Know

Define or describe the following key term from this lesson.

solstice

_____
_____

Academic Vocabulary

Circle the letter of the word that has the closest meaning to the underlined word from this lesson.

1. Earth moves around the sun with its axis tilted, so that <u>eventually</u> the sun's rays will directly strike the latitude 23 1/2°.

eventually

   a. instantly   **b.** finally   **c.** probably

2. The <u>cycle</u> of the seasons repeats itself each year as Earth revolves around the sun.

cycle

   a. series   **b.** age   **c.** axis

The Poles *(pages 57–58)*

Evaluating

As you read, assess the quality of the information provided.

1. In the quote by Keith Nyitray, are opinions presented as facts? How do you know?

_____

_____

_____

2. Is Nyitray qualified to write on this subject? How do you know?

_____

_____

_____

Terms To Review

Use the following term that you studied earlier in a sentence that reflects the term's meaning.

varies
(Chapter 2, Section 3)

_____

_____

The Greenhouse Effect *(page 58)*

Interpreting

Think about what you already know about global warming. Ask yourself, **How might global warming affect the area where I live?**

_____

_____

_____

_____

 Notes

Terms To Know

Define or describe the following key terms from this lesson.

greenhouse effect >

_____

_____

global warming >

_____

_____

Section Wrap-up

Now that you have read the section, write the answers to the questions that were included in Setting a Purpose for Reading *at the beginning of the lesson.*

How does Earth's position in relation to the sun affect temperatures on Earth?

_____

_____

How does Earth's rotation cause day and night?

_____

_____

What is Earth's position in relation to the sun during each season?

_____

_____

How might global warming affect Earth's air, land, and water?

_____

_____

Chapter 3, Section 2
Factors Affecting Climate
(Pages 59–64)

Reason To Read

Setting a Purpose for Reading Think about these questions as you read:
- How do latitude and elevation affect climate?
- What role do wind patterns and ocean currents play in Earth's climates?
- How do landforms and climate patterns influence each other?

Main Idea

As you read pages 59–64 in your textbook, complete this graphic organizer by listing factors that cause both winds and ocean currents.

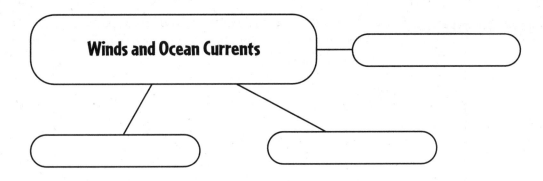

Latitude and Climate (pages 59–61)

Analyzing

As you read this section, think about its organization and main ideas. Then write a sentence explaining the organization and list three main ideas.

Organization

_____

_____

Main Ideas

_____

_____

_____

_____

Places To Locate

Write the letter of the correct location in Group B next to the correct place from this lesson in Group A. One location will not be used.

Group A

_____ **1.** low latitudes

_____ **2.** high latitudes

_____ **3.** mid-latitudes

_____ **4.** Arctic Circle

_____ **5.** Antarctic Circle

Group B

a. latitude 66°N

b. between the Tropic of Cancer and the Arctic Circle and between the Tropic of Capricorn and the Antarctic Circle

c. on the Equator

d. between the Tropic of Cancer and the Tropic of Capricorn

e. the polar area north of the Arctic Circle and the polar area south of the Antarctic Circle

f. latitude 66°S

Academic Vocabulary

Define the following academic vocabulary words from this lesson.

annual

_____

_____

 Key Points

 Notes

portions

Terms To Review

Use the following term that you studied earlier in a sentence that reflects the term's meaning.

revolution
(Chapter 3, Section 1)

Elevation and Climate (page 61)

Skimming

Read the title and quickly look over the section to get a general idea of the section's content. Next, write a sentence or two that explains what the section is about.

Academic Vocabulary

Define the following academic vocabulary word from this lesson.

retain

Wind and Ocean Currents (pages 61–63)

Outlining

As you read, complete the following sentences. Doing so will help you summarize the lesson.

1. Winds occur because the _____ heats up the earth's atmosphere and surface unevenly.

2. Because Earth rotates to the east, the global winds are displaced

   _____ in the Northern Hemisphere and _____ in the Southern Hemisphere.

3. Driven by temperature, _____ creates precipitation.

Terms To Know

Fill in the blank with the correct term from this lesson from the list below.

> Coriolis effect

> doldrums

> currents

1. The _____ is a narrow, generally windless band located near the Equator.

2. The _____ is also observed in ocean _____ , causing them to move in clockwise circles in the Northern Hemisphere and counterclockwise circles in the Southern Hemisphere.

3. Cold or warm streams of seawater that flow in the oceans, generally in a circular pattern are called _____ .

Academic Vocabulary

Fill in the blank with the correct term from this lesson from the list below.

normally   displaced   perspective

1. In an El Niño year, the _____ low atmospheric pressure over the western Pacific Ocean rises.

2. Because Earth rotates to the east, the global winds are _____ , or shifted clockwise in the Northern Hemisphere.

Landforms and Climate (page 63–64)

Predicting

Read the title and main headings of the lesson. Write a statement predicting what the lesson will be about and what will be included in the text.

_____

_____

_____

 **Key Points**

 **Notes**

Terms To Know

Write the letter of the correct definition in Group B next to the correct term in Group A. One definition will not be used.

Group A

_____ **1.** windward

_____ **2.** leeward

_____ **3.** rain shadow

Group B

a. facing away from the direction from which the wind is blowing

b. dry area found on the side of the mountain that is facing away from the direction the wind is blowing

c. moisture that falls on mountains in the form of rain and snow

d. facing toward the direction from which the wind is blowing

Section Wrap-up

Now that you have read the section, write the answers to the questions that were included in Setting a Purpose for Reading *at the beginning of the lesson.*

How do latitude and elevation affect climate?

_____

_____

What role do wind patterns and ocean currents play in Earth's climates?

_____

_____

How do landforms and climate patterns influence each other?

_____

_____

Chapter 3, Section 3
World Climate Patterns

(Pages 65–69)

Reason To Read

Setting a Purpose for Reading Think about these questions as you read:
- How do geographers classify the climate regions of the world?
- Which kinds of vegetation are characteristic of each climate region?
- How do recurring phenomena influence climate patterns?

Main Idea

As you read pages 65–69 in your textbook, complete this graphic organizer by filling in a brief description of each climate region.

Climate	Description
Tropical	
Dry	
Mid-Latitude	
High Latitude	
Highlands	

Climate Regions (pages 65–69)

Connecting

As you read, compare the climate regions with the climate in the area where you live. Identify the climate region where you live. Summarize the climate and vegetation of the area where you live in a paragraph.

_____

_____

_____

_____

Terms To Know

Write the letter of the correct definition in Group B next to the correct place from this lesson in Group A. One definition will not be used.

Group A

_____ **1.** natural vegetation

_____ **2.** oasis

_____ **3.** coniferous

_____ **4.** deciduous

_____ **5.** mixed forests

_____ **6.** chaparral

_____ **7.** prairies

_____ **8.** permafrost

Group B

a. forests with coniferous and deciduous trees

b. trees with leaves that change color and drop in autumn

c. thickets of woody bushes and short trees

d. cone-bearing trees

e. the plant life that grows in an area where the natural environment is unchanged by human activity

f. permanently frozen subsoil

g. scrub and cactus plants

h. inland grasslands

i. a small area in the desert where water and vegetation are found

Places To Locate

Explain why the following places from this lesson are important.

Tropics

_____

_____

 **Key Points**

 **Notes**

Mediterranean Sea >

_____

_____

Academic Vocabulary

Define the following academic vocabulary words from this lesson.

layers >

_____

_____

_____

consists >

_____

_____

Climatic Changes (page 69)

Interpreting

Think about human interaction with the environment and how it affects climate. Ask yourself, How might this human interaction affect the environment where I live?

_____

_____

Terms To Know

Define or describe the following key terms from this lesson.

hypothesis >

_____

_____

_____

smog >

_____

_____

Academic Vocabulary

Define the following academic vocabulary words from this lesson.

output

_____
_____
_____

visible

_____
_____

Terms To Review

Use the following term that you studied earlier in a sentence that reflects the term's meaning.

energy
(Chapter 2, Section 2)

_____
_____

Section Wrap-up

Now that you have read the section, write the answers to the questions that were included in Setting a Purpose for Reading *at the beginning of the lesson.*

How do geographers classify the climate regions of the world?

_____
_____

Which kinds of vegetation are characteristic of each climate region?

_____
_____

How do recurring phenomena influence climate patterns?

_____
_____

Chapter 4, Section 1
World Population

(Pages 75–79)

Reason To Read

Setting a Purpose for Reading Think about these questions as you read:
- What factors affect a country's population growth rate?
- What challenges does population growth pose for the planet?
- Why is the world's population unevenly distributed?

Main Idea

As you read pages 75–79 in your textbook, complete this graphic organizer by listing the challenges created by population growth.

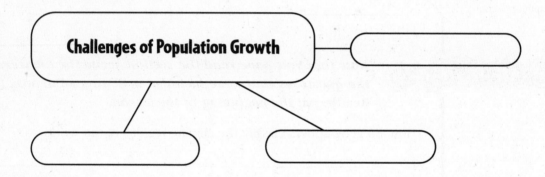

Challenges of Population Growth

Population Growth (pages 75–77)

Drawing Conclusions

As you read, write three details about population growth. Then write a general statement based on these details.

Details

_____
_____
_____
_____

General Statement

_____
_____

Terms To Know

Use each of the following terms from this lesson in a sentence that reflects the term's meaning.

death rate

_____
_____

birthrate

_____
_____

natural increase

_____
_____

doubling time

_____
_____

Places To Locate

Explain why each of the following places from this lesson is important.

Hungary

_____
_____

Germany

_____
_____

Key Points

Notes

Academic Vocabulary

Circle the letter of the word or phrase that has the closest meaning to the underlined word from this lesson.

> **declined**

1. Death rates have <u>declined</u> faster than birth rates.

   a. improved   **b.** decreased   **c.** leveled

> **resources**

2. The faster the population grows, the faster the world's natural <u>resources</u> are used up.

   a. favors   **b.** goals   **c.** materials

Terms To Review

Use each of the following terms that you studied earlier in a sentence that reflects the term's meaning.

> **available**
> (Chapter 2, Section 3)

_____

_____

_____

> **annual**
> (Chapter 3, Section 2)

_____

_____

Population Distribution (pages 77–79)

Clarifying

As you read this section, write down terms or concepts you find confusing. Then go back and reread the section to clarify the confusing parts. Write an explanation of the terms and concepts. If you are still confused, write your questions down and ask your teacher to clarify.

_____

_____

_____

Terms To Know

Write the letter of the correct definition in Group B next to the correct term from this lesson in Group A. One definition will not be used.

Group A

____ **1.** population distribution

____ **2.** population density

____ **3.** migration

Group B

a. the average number of people living on a square mile or kilometer of land

b. the movement of people from place to place

c. the kind of human activity in a region

d. the pattern of human settlement

Places To Locate

Explain why each of the following places from this lesson is important.

Bangladesh ⟩ _____
_____
_____

Mexico City ⟩ _____
_____

Academic Vocabulary

Define the following academic vocabulary words from this lesson.

concentrated ⟩ _____
_____

overall ⟩ _____
_____

Terms To Review

Use the following term that you studied earlier in a sentence that reflects the term's meaning.

movement
(Chapter 1, Section 1) ⟩ _____
_____

Key Points

Notes

Section Wrap-up

Now that you have read the section, write the answers to the questions that were included in Setting a Purpose for Reading *at the beginning of the lesson.*

What factors affect a country's population growth rate?

_____

_____

_____

_____

What challenges does population growth pose for the planet?

_____

_____

_____

_____

Why is the world's population unevenly distributed?

_____

_____

_____

_____

Chapter 4, Section 2
Global Cultures

(Pages 80–85)

Reason To Read

Setting a Purpose for Reading Think about these questions as you read:
- What factors define a culture?
- What are the major culture regions of the world?
- What developments have affected interaction between cultures in recent years?

Main Idea

As you read pages 80–85 in your textbook, complete this graphic organizer by listing the world culture regions.

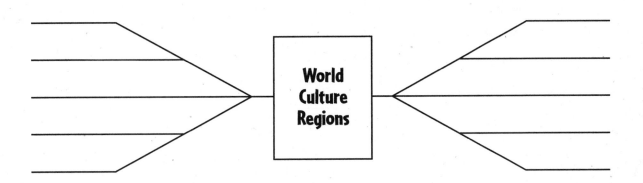

Elements of Culture (pages 80–84)

Reviewing

As you read the lesson, fill in the chart below. Review your chart when you have finished reading and revise as needed.

Elements of Culture	Functions of Each Element

Terms To Know

Define or describe the following key terms from this lesson.

culture

_____

_____

language families

_____

_____

ethnic group

_____

_____

culture region

_____

_____

Academic Vocabulary

Fill in the blank with the correct term from this lesson from the list below.

communicate   fluctuate   factors

1. Two _____ that geographers study to understand a culture are religion and government.

2. Speaking a common language is one way people _____ information and experiences with other people.

Terms To Review

Use each of the following terms that you studied earlier in a sentence that reflects the term's meaning.

elements
(Chapter 2, Section 2)

_____

_____

aspects
(Chapter 3, Section 1)

_____

_____

Cultural Change *(pages 84–85)*

Previewing

Preview the section to get an idea of what is ahead. First, skim the section. Then write a sentence or two explaining what you think you will be learning. After you have finished reading, revise your statements as necessary.

_____

_____

_____

Terms To Know

Define or describe the following key terms from this lesson.

cultural diffusion

_____

_____

culture hearths >

_____
_____

Places To Locate

Explain why the following places from this lesson are important.

Egypt >

_____
_____

Iraq >

_____
_____

Pakistan >

_____
_____

China >

_____
_____

Mexico >

_____
_____

Academic Vocabulary

Define the following academic vocabulary words from this lesson.

emerged >

_____
_____

migrate >

_____
_____

 *Now that you have read the section, write the answers to the questions that were included in* Setting a Purpose for Reading *at the beginning of the lesson.*

What factors define a culture?

_____

_____

_____

_____

What are the major culture regions of the world?

_____

_____

_____

_____

What developments have affected interaction between cultures in recent years?

_____

_____

_____

_____

Chapter 4, Section 3
Political and Economic Systems

(Pages 86–90)

Reason To Read

Setting a Purpose for Reading Think about these questions as you read:
- What are the various levels of government?
- What are the major types of governments in the world today?
- What are the major types of economic systems in the world?

Main Idea

As you read pages 86–90 in your textbook, complete this outline.

Political and Economic Systems

I. Levels of government

    A. _____

    B. _____

II. Types of Government

    A. _____

    B. _____

    C. _____

III. Economic Systems

    A. _____

    B. _____

    C. _____

    D. _____

Features of Government (page 86)

Questioning

As you read, write a question about the main ideas presented in the text. After you have finished reading, write the answer to this question.

_____

_____

Academic Vocabulary

Define the following academic vocabulary words from this lesson.

enforce

_____

_____

defined

_____

_____

Levels of Government (page 87)

Synthesizing

As you read, think about the main ideas of the section. Ask yourself, How do the governments of the United Kingdom and the United States differ?

_____

_____

Terms To Know

Use the following terms from this lesson in a sentence that reflects the term's meaning.

unitary system

_____

_____

federal system

_____

_____

Types of Government *(pages 87–89)*

Previewing

Preview the section to get an idea of what is ahead. Skim the section, then write a sentence or two explaining the lesson content. After you have finished reading, revise your statements as necessary.

_____

_____

_____

Terms To Know

Write the letter of the correct definition in Group B next to the correct term from this lesson in Group A. One definition will not be used.

Group A

____ **1.** autocracy

____ **2.** oligarchy

____ **3.** democracy

Group B

a. rule belongs to a single person

b. independent territories have power

c. small group holds power

d. leaders rule with the consent of the citizens

Academic Vocabulary

Fill in the blanks with the correct terms from this lesson from the list below. One term will not be used.

colleague   individual   authority

1. In a totalitarian dictatorship, the _____ to rule is

given to an _____ leader.

Economic Systems (pages 89–90)

Scanning

Quickly scan over the lines of text, look for key words or phrases that will tell you what the section is about. Write the key words or phrases. Then use the key words and phrases to write a statement explaining what the section is about.

Key words or phrases >

_____

_____

_____

What this section is about >

_____

_____

_____

_____

Terms To Know

Write the letter of the correct definition in Group B next to the correct term from this lesson in Group A. One definition will not be used.

Group A

____ **1.** traditional economy

____ **2.** market economy

____ **3.** mixed economy

____ **4.** command economy

Group B

a. the government supports and regulates free enterprise

b. companies employ people who work in their homes.

c. the government directs the means of production and distribution

d. habit and custom determine the rules for all economic activity

e. individuals and private groups decide what to produce

Academic Vocabulary

 tradition

Circle the letter of the word that has the closest meaning to the underlined word from this lesson.

1. In Inuit society, it was a <u>tradition</u> to share the animals caught during a hunt with other families in the village.

    a. commercial trade   **b.** legal agreement   **c.** common practice

survive

2. The Inuit society was able to <u>survive</u> the Arctic climate because its members helped each other meet their needs.

    a. exist   **b.** change   **c.** increase

Section Wrap-up

Now that you have read the section, write the answers to the questions that were included in Setting a Purpose for Reading *at the beginning of the lesson.*

What are the various levels of government?

_____

_____

What are the major types of governments in the world today?

_____

_____

What are the major types of economic systems in the world?

_____

_____

Chapter 4, Section 4
Resources, Trade, and the Environment

(Pages 91–95)

Reason To Read

Setting a Purpose for Reading Think about these questions as you read:
- What types of energy most likely will be used in societies in the future?
- What factors determine a country's economic development and trade relationships?
- How do human economic activities affect the environment?

Main Idea

As you read pages 91–95 in your textbook, complete this graphic organizer by listing types of renewable energy resources.

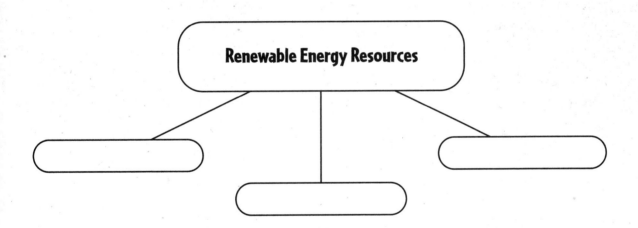

Resources *(pages 91–92)*

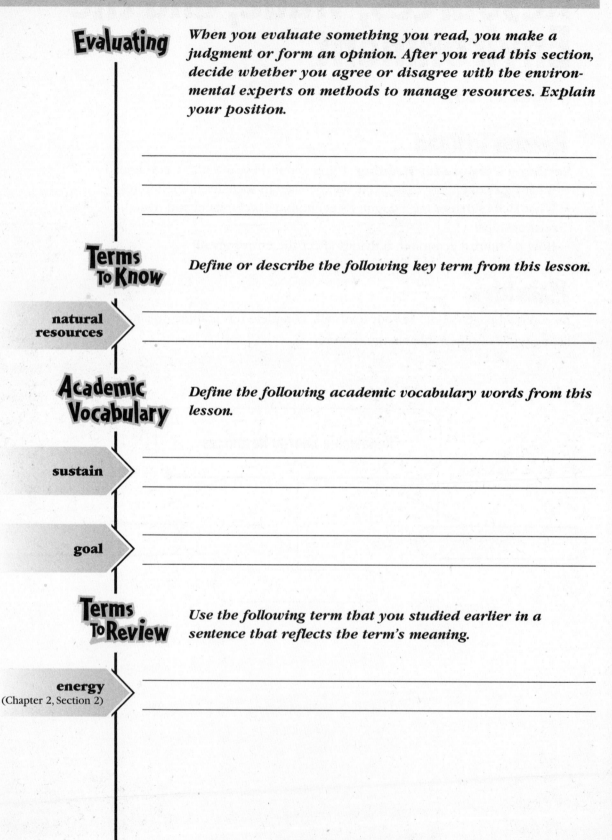

Evaluating

When you evaluate something you read, you make a judgment or form an opinion. After you read this section, decide whether you agree or disagree with the environmental experts on methods to manage resources. Explain your position.

_____

_____

_____

Terms To Know

Define or describe the following key term from this lesson.

natural resources

_____

_____

Academic Vocabulary

Define the following academic vocabulary words from this lesson.

sustain

_____

_____

goal

_____

Terms To Review

Use the following term that you studied earlier in a sentence that reflects the term's meaning.

energy
(Chapter 2, Section 2)

_____

_____

Chapter 4, Section 4

Economic Development (page 93)

Analyzing

As you read this section, think about its organization and main ideas. Then write a sentence explaining the organization and list the main ideas.

Organization
_____
_____

Main Ideas
_____
_____
_____
_____

Terms To Know

Define or describe the following key terms from this lesson.

developed countries
_____
_____

developing countries
_____
_____
_____

industrialization
_____
_____

Academic Vocabulary

Define the following academic vocabulary words from this lesson.

involve
_____
_____

acquiring
_____
_____

Terms To Review

Use each of the following terms that you studied earlier in a sentence that reflects the term's meaning.

economy
(Chapter 1, Section 2)

_____

_____

promote
(Chapter 3, Section 1)

_____

_____

World Trade *(pages 93–94)*

Skimming

Read the title and quickly look over the section to get a general idea of the section's content. Then write a sentence or two explaining what the section is about.

_____

_____

Terms To Know

Define or describe the following key term from this lesson.

free trade

_____

_____

Academic Vocabulary

Fill in the blank with the correct term from this lesson from the list below.

export   locate   equate

1. Countries _____ their specialized products to other countries that cannot produce those goods.

2. Multinational companies often _____ their assembly operations in developing countries with low labor costs.

People and the Environment *(pages 94–95)*

Summarizing

As you read, complete the following sentences. Doing so will help you summarize the section.

1. A major environmental challenge today is air, water, and land

   _____ .

2. When fertilizers and pesticides from farms seep into the ground, the

   _____ _____ can become polluted.

3. The burning of _____ - _____ by

   vehicles and _____ is the main source of air pollution.

4. What harms one part of a natural _____ harms all other

   parts because the earth's land, air, and water are _____ .

Terms To Know

Define or describe the following key term from this lesson.

pollution

_____

_____

Academic Vocabulary

Circle the letter of the word that has the closest meaning to the underlined word.

conduct

1. The leaves of trees <u>conduct</u> photosynthesis.

   a. set up   **b.** go through   **c.** prevent

process

2. Photosynthesis is the <u>process</u> by which plants take in carbon dioxide and, in the presence of sunlight, produce carbohydrates.

   a. method   **b.** industry   **c.** thought

Terms To Review

Use each of the following terms that you studied earlier in a sentence that reflects the term's meaning.

cycle
(Chapter 3, Section 1)

_____

_____

_____

global warming
(Chapter 3, Section 1)

_____

_____

_____

Section Wrap-up

Now that you have read the section, write the answers to the questions that were included in Setting a Purpose for Reading *at the beginning of the lesson.*

What types of energy most likely will be used in societies in the future?

_____

_____

What factors determine a country's economic development and trade relationships?

_____

_____

How do human economic activities affect the environment?

_____

_____

Chapter 5, Section 1
The Land
(Pages 115–120)

Reason To Read

Setting a Purpose for Reading Think about these questions as you read:
- What are some key similarities and differences in the physical geography of the United States and Canada?
- Why have rivers played such an important role in this region's development?
- What geographic factors have made the United States and Canada so rich in natural resources?

Main Idea

As you read pages 115–120 in your textbook, complete this graphic organizer by listing the major minerals found in the United States and Canada.

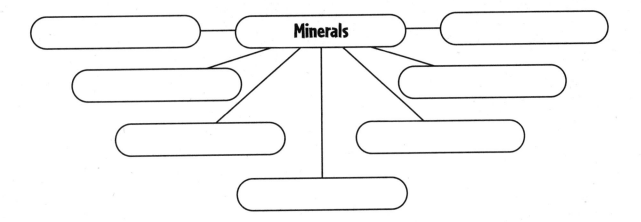

The Land *(pages 115–120)*

Interpreting

Think about human interaction with the environment and how it helps determine settlement patterns. Ask yourelf, How have the physical geography and natural resources of the United States and Canada affected human settlement?

_____

_____

_____

_____

Academic Vocabulary

Define the following academic vocabulary words from this lesson.

enormous

_____

_____

displays

_____

_____

Terms To Review

Use the following term that you studied earlier in a sentence that reflects the term's meaning from this lesson.

emerging
(Chapter 4, Section 2)

_____

_____

 **Key Points**

 **Notes**

A Fortune in Water *(page 116–119)*

Evaluating   *As you read, form an opinion about the following information.*

1. In the quote by Barbara Kingsolver, are opinions being presented as facts?

_____

_____

2. Is Kingsolver qualified to write on this subject? How do you know?

_____

_____

Terms To Know   *Write the letter of the correct definition in Group B next to the correct term from this lesson in Group A. One definition will not be used.*

Group A

_____ **1.** divide

_____ **2.** headwaters

_____ **3.** tributaries

_____ **4.** fall line

Group B

a. source

b. freshwater lakes and rivers

c. a high point or ridge that determines the direction that rivers flow

d. a boundary that marks the place where the higher land of the Piedmont region drops to the lower Atlantic Coastal Plain

e. brooks, rivers, and streams

Academic Vocabulary   *Circle the letter of the word or phrase that has the closest meaning to the underlined word from this lesson.*

link

1. The St. Lawrence Seaway provides a <u>link</u> between the Atlantic Ocean and the Great Lakes.

   a. break   **b.** connection   **c.** border

prime

2. The Mississippi River is the <u>prime</u> mover of goods across the United States.

   a. unusable   **b.** aging   **c.** main

Key Points

Terms To Review

source
(Chapter 2, Section 3)

glacier
(Chapter 2, Section 2)

Natural Resources *(page 119–120)*

Predicting

Terms To Know

fisheries

Academic Vocabulary

access

Notes

Use the follownig terms that you studied earlier in a sentence that reflects the term's meaning from this lesson.

_____

_____

_____

_____

Read the title and main headings of the lesson. Write a statement predicting what the lesson will cover and what will be included in the text. As you read, change your prediction if it does not match what you learn.

_____

_____

_____

Define or describe the following key term from this lesson.

_____

_____

Define the following academic vocabulary word from this lesson.

_____

_____

Terms To Review

Use the following term that you studied earlier in a sentence that reflects the term's meaning from this lesson.

economy
(Chapter 1, Section 2)

_____

_____

Section Wrap-up

Now that you have read the section, write the answers to the questions that were included in **Setting a Purpose for Reading** *at the beginning of the lesson.*

What are some key similarities and differences in the physical geography of the United States and Canada?

_____

_____

Why have rivers played such an important role in this region's development?

_____

_____

_____

What geographic factors have made the United States and Canada so rich in natural resources?

_____

_____

Chapter 5, Section 2
Climate and Vegetation

(Pages 121–125)

Reason To Read

Setting a Purpose for Reading Think about these questions as you read:

- Which climate zones are found in the United States and Canada?
- In what ways do winds, ocean currents, latitude, and landforms affect the region's climates?
- What kinds of weather hazards affect the United States and Canada?
- How has human settlement affected the natural vegetation of the United States and Canada?

Main Idea

As you read pages 121–125 in your textbook, complete this graphic organizer by listing the factors that contribute to the varying climate and vegetation found in the western climates of the United States and Canada.

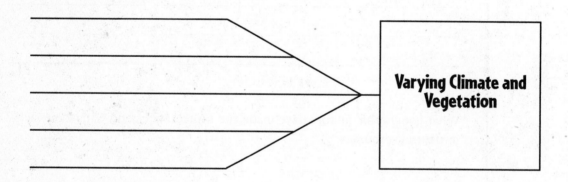

Varying Climate and Vegetation

 Key Points

 Notes

A Varied Region *(page 121)*

Interpreting

Think about the latitudes of the United States and Canada. Ask yourself, why does Canada have a much colder climate than most of the United States?

_____

_____

Academic Vocabulary

Define the following academic vocabulary word from this lesson.

brief

_____

_____

Northern Climates *(page 122)*

Visualizing

Visualize the information described is this section to help you understand and remember what you have read. First, read the section. Next, ask yourself, What would this look like? Finally, write a description of the pictures you visualized on the lines below.

_____

_____

_____

Terms To Review

Use each of the following terms that you studied earlier in a sentence that reflects the term's meaning from this lesson.

coniferous
(Chapter 3, Section 3)

_____

_____

survive
(Chapter 4, Section 3)

_____

Western Climates (pages 122–124)

Reviewing

Think about the northern and western climates. Explain the differences between these climates.

_____

_____

Terms To Know

Fill in the blank with the correct term from this lesson listed below. One term will be used twice.

chinook

1. The _____ is the elevation above which trees cannot grow.

2. The _____ is a warm, dry wind that blows down the eastern slopes of the Rockies in late winter and early spring.

timberline

3. Only lichens and mosses can be found above the _____ .

Academic Vocabulary

Define the following academic vocabulary word from this lesson.

releases

_____

_____

Terms To Review

Use each of the following terms that you studied earlier in a sentence that reflects the term's meaning from this lesson.

currents
(Chapter 3, Section 2)

_____

_____

rain shadow
(Chapter 3, Section 2)

_____

_____

Interior Climates (pages 124–125)

Questioning

As you read, write two questions about the main ideas presented in the text. After you have finished reading, write the answers to these questions.

1. _____

_____

_____

2. _____

_____

_____

Terms To Know

Define or describe the following key terms from this lesson.

prairies _____

_____

supercells _____

Academic Vocabulary

Circle the letter of the word that has the closest meaning to the underlined word from this lesson.

restored

1. Improved ways of farming and conservation have <u>restored</u> the soil of the Great Plains.

   a. revived   **b.** ruined   **c.** recorded

eroded

2. The dry weather and the wind <u>eroded</u> the loose soil and turned the Great Plains into an empty wasteland.

   a. built up   **b.** wore away   **c.** passed over

Eastern Climates (page 125)

Responding

As you read this section, think about what attracts your attention as you read. Write down facts you find interesting or surprising in the lesson.

_____

_____

_____

_____

Terms To Know

Define or describe the following key terms from this lesson.

hurricanes

_____

_____

blizzards

_____

_____

Terms To Review

Use the following term that you studied earlier in a sentence that reflects the term's meaning from this lesson.

deciduous
(Chapter 3, Section 3)

_____

_____

Tropical Climates (page 125)

Previewing

Preview the section to get an idea of what is ahead. First, skim the section. Then write a sentence or two predicting what you think you will be learning. After you have finished reading, revise your prediction as necessary.

_____

_____

_____

Now that you have read the section, write the answers to the questions that were included in Setting a Purpose for Reading *at the beginning of the lesson.*

Which climate zones are found in the United States and Canada?

_____

_____

In what ways do winds, ocean currents, latitude, and landforms affect the region's climates?

_____

_____

What kinds of weather hazards affect the United States and Canada?

_____

_____

How has human settlement affected the natural vegetation of the United States and Canada?

_____

_____

Chapter 6, Section 1
Population Patterns

(Pages 133–137)

Reason To Read

Setting a Purpose for Reading Think about these questions as you read:

- Who are the peoples of the United States and Canada?
- How are population patterns in the United States and Canada influenced by the region's physical geography?
- What geographic factors encouraged the industrialization and urbanization of the United States and Canada?

Main Idea

As you read pages 133–137 in your textbook, complete this graphic organizer by listing the cities that comprise the megalopolis Boswash.

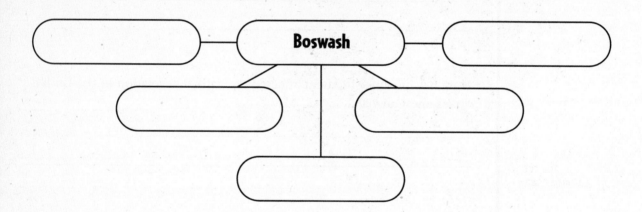

The People (pages 133–134)

Inferring

As you read the lesson, look for clues in the descriptions and events that might help you draw a conclusion. To answer the following question, first identify the descriptions and events that help you answer the question, then draw a conclusion.

Why did immigrants refer to the United States as the land "where the streets are paved with gold"?

Descriptions and events >

_____

_____

Conclusion >

_____

_____

Terms To Know

Define or describe the following key terms from this lesson.

immigration >

_____

_____

Native American >

_____

_____

Academic Vocabulary

Define the following vocabulary words from this lesson.

diverse >

_____

_____

seek >

_____

_____

Terms To Review

Use each of the following terms that you studied earlier in a sentence that reflects the term's meaning from this lesson.

region
(Chapter 1, Section 1)

_____
_____
_____

benefited
(Chapter 3, Section 2)

_____
_____

Population Density (pages 134–135)

Determining the Main Idea

As you read, write the main idea of the lesson. Review your statement when you have finished reading and revise as needed.

_____
_____

Terms To Know

Define or describe the following key term from this lesson.

Sunbelt

_____
_____

The Cities (pages 135–137)

Connecting

As you read, compare the places to live in the United States and Canada. Describe the area where you live.

_____
_____
_____
_____
_____

Terms To Know

Write the letter of the correct definition in Group B next to the correct term from this lesson in Group A. One definition will not be used.

Group A

____ **1.** urbanization

____ **2.** metropolitan area

____ **3.** suburbs

____ **4.** megalopolis

____ **5.** mobility

Group B

a. source

b. freshwater lakes and rivers

c. a high point or ridge that determines the direction that rivers flow

d. a boundary that marks the place where the higher land of the Piedmont region drops to the lower Atlantic Coastal Plain

e. brooks, rivers, and streams

Academic Vocabulary

Circle the letter of the word or phrase that has the closest meaning to the underlined word from this lesson.

vision

1. The <u>vision</u> of Pierre L'Enfant led to the planned city of Washington, D.C.

   a. foresight   **b.** look   **c.** travels

design

2. Aerospace industries <u>design</u> and produce airplanes, satellites, and space vehicles.

   a. obtain   **b.** improve   **c.** plan

Terms To Review

Use each of the following terms that you studied earlier in a sentence that reflects the term's meaning from this lesson.

birthrate
(Chapter 4, Section 1)

_____

_____

locations
(Chapter 1, Section 1)

_____

_____

Section Wrap-up

Now that you have read the section, write the answers to the questions that were included in Setting a Purpose for Reading *at the beginning of the lesson.*

Who are the peoples of the United States and Canada?

_____

_____

_____

How are population patterns in the United States and Canada influenced by the region's physical geography?

_____

_____

_____

What geographic factors encouraged the industrialization and urbanization of the United States and Canada?

_____

_____

_____

Chapter 6, Section 2
History and Government

(Pages 140-145)

Reason To Read

Setting a Purpose for Reading Think about these questions as you read:
- What was life like for the earliest Americans and for European settlers?
- How did industrialization and technology enable westward expansion in North America?
- How do the governments of the United States and Canada differ?

Main Idea

As you read pages 140-145 in your textbook, complete this graphic organizer by filling in the results of each innovation on farming the Great Plains.

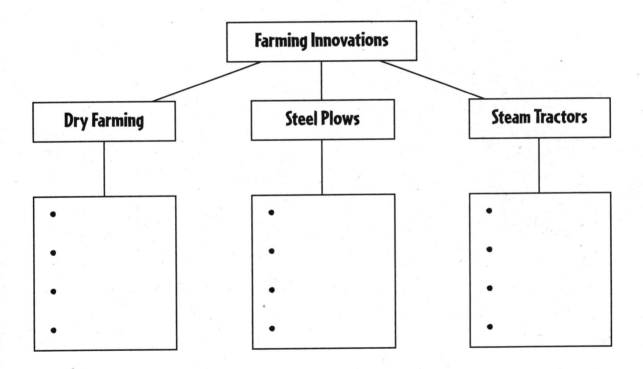

History (pages 140–144)

Sequencing

As you read, number the following events in the history of the United States and Canada in the correct order.

_____ France was forced to give up much of its North American empire to Great Britain.

_____ The transcontinental railroads were completed.

_____ Nomads crossed a land bridge from Asia to Alaska.

_____ European migration began.

_____ Industrialization began in North America.

Terms To Know

Define or describe the following key terms from this lesson.

republic _____

_____

_____

Underground Railroad _____

_____

_____

dry farming _____

_____

Places To Locate

Fill in the blank with the correct place from this lesson from the list below.

Texas	Hudson Bay	Ohio	Northwest Territories
Nunavut	Hawaii	Ontario	Quebec
Alaska	Pennsylvania	New Brunswick	Nova Scotia
Yukon Territory			

1. The United States purchased _____ from Russia in 1867.

2. The _____ is an inland sea in east-central Canada.

3. _____ was a former Mexican territory that became an independent republic in 1836 and then joined the United States in 1845.

4. In 1867, the four colonies of _____, _____, _____, and _____ united as the Dominion of Canada.

5. The United States acquired the Pacific islands called _____ .

6. The two northeastern states of _____ and _____ had large supplies of coal that were used to power steam engines.

7. Canada's northern territories are _____ , _____ , and _____ .

Academic Vocabulary

Define the following academic vocabulary words from this lesson.

theory >
_____
_____
_____

contrast >
_____
_____

Terms To Review

Use the following terms that you studied earlier in a sentence that reflects the term's meaning from this lesson.

expanded
(Chapter 2, Section 2) >
_____
_____

industrialization
(Chapter 4, Section 4) >
_____
_____

Government *(pages 144–145)*

Monitoring Comprehension

As you read the lesson, write down questions you have about what you read. When you have finished reading the lesson, answer your questions.

_____

_____

_____

_____

Terms To Know

Fill in the blank with the correct term from this lesson from the list below. One term will not be used.

cabinet	Parliament	amendments
Constitution	dominion	Bill of Rights

1. Canada was created as a _____ , which is a partially self-governing country with close ties to Great Britain.

2. The national legislature of Canada is called _____ .

3. The United States' plan of government is called the _____ .

4. The _____ guarantees such freedoms as religion, speech, and press.

5. Changes in the Constitution are called _____ .

Academic Vocabulary

Define the following academic vocabulary words from this lesson.

guarantee ＞ _____

_____

drafted ＞ _____

_____

Terms To Review

Use each of these terms that you studied earlier in a sentence that reflects the term's meaning from this lesson.

federal system
(Chapter 4, Section 3)

_____

_____

democracies
(Chapter 4, Section 3)

_____

_____

Section Wrap-up

Now that you have read the section, write the answers to the questions that were included in Setting a Purpose for Reading *at the beginning of the lesson.*

What was life like for the earliest Americans and for European settlers?

_____

_____

_____

_____

How did industrialization and technology enable westward expansion in North America?

_____

_____

_____

_____

How do the governments of the United States and Canada differ?

_____

_____

_____

_____

Chapter 6, Section 3
Cultures and Lifestyles

(Pages 146–151)

Reason To Read

Setting a Purpose for Reading Think about these questions as you read:
- How do the religious practices and languages of the region reflect the immigrant history of the United States and Canada?
- How do the arts of the United States and Canada reflect the region's colonial past?
- What kinds of educational and health care systems serve the people of the region?

Main Idea

As you read pages 146–151 in your textbook, complete this graphic organizer by listing authors and topics they wrote about.

Author	Topic
James Fenimore Cooper	Life in North America

Cultural Characteristics (pages 146–148)

Drawing Conclusions

As you read, write three details about the cultural characteristics of the United States and Canada. Then write a conclusion based on these details.

Details

_____
_____
_____
_____

Conclusion

_____
_____

Terms To Know

Define or describe the following key term from this lesson.

bilingual

_____
_____

Academic Vocabulary

Define the following academic vocabulary words from this lesson.

ethnic

_____
_____
_____

evident

_____
_____

Terms To Review

Use the following term that you studied earlier in a sentence that reflects the term's meaning from this lesson.

immigration
(Chapter 6, Section 1)

_____
_____

The Arts *(pages 148–149)*

Responding

As you read this section, think about what attracts your attention as you read. Write down facts you find interesting or surprising in the lesson.

_____

_____

_____

_____

Terms To Know

Define or describe the following key term from this lesson.

jazz

_____

_____

Academic Vocabulary

Define the following academic vocabulary words from this lesson.

dynamic

_____

_____

attitudes

_____

_____

Terms To Review

Use the following term that you studied earlier in a sentence that reflects the term's meaning from this lesson.

traditions
(Chapter 4, Section 3)

_____

_____

Lifestyles (pages 150–151)

Connecting

As you read, think about the lifestyles of Canadians and people in the United States.

Write a paragraph describing your lifestyle.

_____

_____

_____

_____

Terms To Know

Fill in the blank with the correct term from this lesson from the list below.

> socioeconomic status

> literacy rate

> patriotism

1. _____ is loyalty to one's country.

2. The _____ , or level of income and education, of the people in the United States and Canada is high.

3. The _____ in both the United States and Canada is very high because both countries maintain compulsory education requirements.

Academic Vocabulary

Circle the letter of the word or phrase that has the closest meaning to the underlined word from this lesson.

1. Both the United States and Canada have high levels of economic development so the governments <u>devote</u> large amounts of resources to health care.

   a. create   **b.** set aside   **c.** represent

2. Many low-income families in the United States are unable to <u>purchase</u> health insurance.

   a. buy   **b.** give up   **c.** split

Now that you have read the section, write the answers to the questions that were included in Setting a Purpose for Reading *at the beginning of the lesson.*

How do the religious practices and languages of the region reflect the immigrant history of the United States and Canada?

_____

_____

How do the arts of the United States and Canada reflect the region's colonial past?

_____

_____

What kinds of educational and health care systems serve the people of the region?

_____

_____

Chapter 7, Section 1
Living in the United States and Canada

(Pages 157–164)

Reason To Read

Setting a Purpose for Reading Think about these questions as you read:
- What are the effects of physical geography on the region's agriculture?
- What kinds of transportation and communications systems does the region have?
- How are the economies of the United States and Canada interdependent with each other and those in other parts of the world?

Main Idea

As you read pages 157–164 in your textbook, complete this graphic organizer by identifying the location of the "belt" regions listed.

Belt Region	Location
Wheat Belt	
Corn Belt	
Rust Belt	
Sun Belt	

Economic Activities (pages 157–159)

 **Analyzing**

As you read this lesson, think about the organization and main ideas of the text. Then write a sentence explaining the organization and list the main ideas.

Organization >
_____
_____

Main Ideas >
_____
_____

Terms To Know

Use each of the following terms in a sentence that reflects the term's meaning from this lesson.

market economy >
_____
_____

post-industrial >
_____
_____

commodity >
_____
_____

Academic Vocabulary

Define the following academic vocabulary words from this lesson.

emphasis >
_____
_____

roles >
_____
_____

Manufacturing and Service Industries (pages 159–161)

Previewing

Preview the lesson to get an idea of what is ahead. First, skim the lesson. Then write a sentence or two explaining what you think you will be learning. After you have finished reading, revise your statements as necessary.

_____

_____

_____

_____

Terms To Know

Define or describe the following key term from this lesson.

retooling

_____

_____

Terms To Review

Use the following term that you studied earlier in a sentence that reflects the term's meaning from this lesson.

export
(Chapter 4, Section 4)

_____

_____

Transportation and Communications (pages 161–162)

Scanning

Scan the lesson before you begin to read. As you glance quickly over the lines of text, look for key words or phrases that tell you what the text will cover. Write the key words or phrases. Then use the key words and phrases to write a statement explaining the lesson content. Revise your statement when you are finished reading the lesson.

Key words or phrases

_____

_____

_____

 **Key Points**

 **Notes**

> **What the lesson is about**

_____

_____

_____

_____

Terms To Know

Define or describe the following key terms from this lesson.

> **pipelines**

_____

_____

_____

> **monopoly**

_____

_____

_____

Trade and Interdependence (pages 162–163)

Drawing Conclusions

As you read the lesson, look for clues in the descriptions and events that might help you draw a conclusion. To answer the following question, first identify the descriptions and events that help you answer the question, then draw a conclusion.

As a result of NAFTA, what has happened to jobs in the United States?

> **Descriptions and events**

_____

_____

_____

> **Conclusion**

_____

_____

Terms To Know

Define or describe the following key terms from this lesson.

> **trade deficit**

_____

_____

tariffs	_____
	_____
	_____
trade surplus	_____
	_____
	_____

United Against Terrorism (pages 163–164)

Predicting

Read the title and main headings of the lesson. Write a statement predicting what the lesson is about and what will be included in the text. As you read, adjust or change your prediction if it does not match what you learn.

_____

_____

_____

Academic Vocabulary

Circle the letter of the word or phrase that has the closest meaning to the underlined word from this lesson.

1. After the terrorist attacks on the United States on September 11, 2001, the United States <u>resolved</u> to unite against terrorism.

   a. fought   **b.** refused   **c.** promised

2. One way Americans <u>responded</u> to the terrorist attacks of September 11, 2001, was to put up flags to show patriotism.

   a. reacted   **b.** said   **c.** overlooked

Terms To Review

Use the following term that you studied earlier in a sentence that reflects the term's meaning from this lesson.

| process (Chapter 4, Section 4) | _____ |
| | _____ |

 **Key Points**

 **Notes**

 **Section Wrap-up**

Now that you have read the section, write the answers to the questions that were included in **Setting a Purpose for Reading** *at the beginning of the lesson.*

What are the effects of physical geography on the region's agriculture?

_____

_____

_____

_____

What kinds of transportation and communications systems does the region have?

_____

_____

_____

_____

How are the economies of the United States and Canada interdependent with each other and those in other parts of the world?

_____

_____

_____

_____

Chapter 7, Section 2
People and Their Environment

(Pages 165–169)

Reason To Read

Setting a Purpose for Reading Think about these questions as you read:
- How are the United States and Canada learning to manage their natural resources more responsibly?
- What are the causes and effects of pollution in the region? How can pollution be prevented?
- What environmental challenges face the United States and Canada in the twenty-first century, both as individual countries and as a region?

Main Idea

As you read pages 165–169 in your textbook, complete this graphic organizer by listing the causes of pollution.

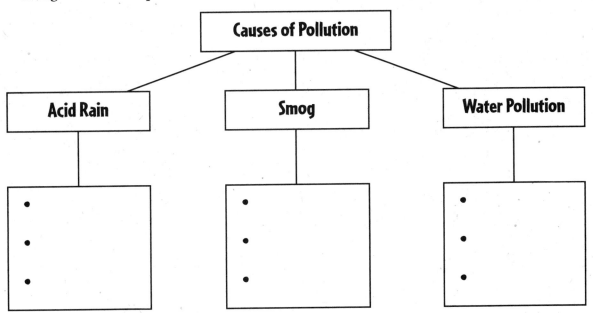

Human Impact (pages 165–166)

Evaluating

When you evaluate something you read, you make a judgment or form an opinion. After you read this section, decide which wildfire policy you most agree with and explain why.

_____

_____

_____

Terms To Know

Define or describe the following key term from this lesson.

clear-cutting

_____

_____

Academic Vocabulary

Define the following academic vocabulary words from this lesson.

impact

_____

_____

evaluating

_____

_____

Pollution (pages 166–167)

Outlining

Complete this outline as you read.

I. Acid Rain

   A. _____

   B. _____

   C. _____

II. Smog

    A. _____

    B. _____

    C. _____

III. Water Pollution

    A. _____

    B. _____

    C. _____

Terms To Know

Define or describe the following key terms from this lesson.

acid rain _____

_____

smog _____

_____

groundwater _____

_____

eutrophication _____

_____

Terms To Review

Use the following term that you studied earlier in a sentence that reflects the term's meaning from this lesson.

metropolitan areas
(Chapter 6, Section 1) _____

_____

Challenge for the Future (page 169)

Summarizing

As you read, complete the following sentences. Doing so will help you summarize the section.

1. One effect of _____ _____ in the Arctic regions of Alaska and Canada is that the _____ _____ is melting at a faster rate than before.

2. As a result of _____ _____ , the permafrost is beginning to _____ . This causes the land to buckle and the foundations of the houses to _____ .

3. Warmer, higher seas lead to an increase in the severity of _____ , causing flooding and _____ .

Academic Vocabulary

Circle the letter of the word or phrase that has the closest meaning to the underlined word from this lesson.

1. An important challenge facing the world is <u>monitoring</u> the effects of global warming.

   a. keeping track of   **b.** giving away   **c.** adding to

2. New Orleans needs to respond <u>appropriately</u> to the effects of global warming so it does not completely flood.

   a. slowly   **b.** carelessly   **c.** correctly

Terms To Review

Use each of these terms that you studied earlier in a sentence that reflects the term's meaning from this lesson.

global warming
(Chapter 3, Section 1)

_____

_____

permafrost
(Chapter 3, Section 3)

_____

_____

Section Wrap-up

Now that you have read the section, write the answers to the questions that were included in Setting a Purpose for Reading *at the beginning of the lesson.*

How are the United States and Canada learning to manage their natural resources more responsibly?

_____

_____

_____

_____

What are the causes and effects of pollution in the region? How can pollution be prevented?

_____

_____

_____

_____

What environmental challenges face the United States and Canada in the twenty-first century, both as individual countries and as a region?

_____

_____

_____

_____

Chapter 8, Section 1
The Land
(Pages 193–198)

Reason To Read

Setting a Purpose for Reading Think about these questions as you read:
- How do geographers divide the large region known as Latin America?
- What factors have shaped Latin America's landforms?
- How has the Latin American landscape influenced patterns of human settlement?
- What natural resources make Latin America an economically important region?

Main Idea

As you read pages 193–198 in your textbook, complete this graphic organizer by listing the countries drained by the Amazon Basin.

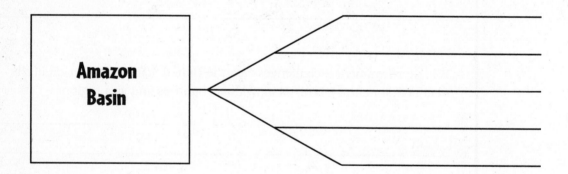

A Vast Region (pages 193–194)

Previewing

Preview the section to get an idea of what is ahead. First, skim the section. Then write a sentence or two explaining what you think you will be learning. After you have finished reading, revise your statements as necessary.

_____

_____

_____

Academic Vocabulary

Define the following academic vocabulary words from this lesson.

region

_____

_____

consist

_____

Terms To Review

Use the following term that you studied earlier in a sentence that reflects the term's meaning.

area
(Chapter 2, Section 1)

_____

_____

Mountains and Plateaus (pages 194–196)

Visualizing

Visualize the information described in this section to help you understand and remember what you have read. First, read the section. Next, ask yourself, What would this look like? Finally, write a description of the pictures you visualized on the lines below.

_____

_____

_____

Key Points

Notes

Terms To Know

Write the letter of the correct definition in Group B next to the correct term in Group A. One definition will not be used.

Group A	Group B
_____ **1.** cordilleras	**a.** vast grasslands used for grazing
	b. high plain; a region in Peru and Bolivia encircled by the Andes peaks
_____ **2.** altiplano	
	c. parallel ranges of mountains
_____ **3.** escarpment	**d.** steep cliff or slope between a higher and lower land surface

Academic Vocabulary

Circle the letter of the word or phrase that has the closest meaning to the underlined word from this lesson.

despite

1. Latin America has been a place of settlement for thousands of years <u>despite</u> its rugged landscape.

    a. even though   **b.** on the other hand   **c.** because of

isolate

2. The mountains sometimes <u>isolate</u> the people of Latin America because they act as barriers to transportation in the region.

    a. anger   **b.** separate   **c.** attract

Terms To Review

Use each of these terms that you studied earlier in a sentence that reflects the term's meaning.

movement
(Chapter 1, Section 1)

_____

_____

range
(Chapter 1, Section 2)

_____

_____

Lowlands and Plains *(page 196–197)*

Determining the Main Idea

As you read, write down the main idea of the passage. Review your statement when you have finished reading and revise as needed.

_____

_____

_____

Terms To Know

Define or describe the following key terms from this lesson.

llanos

_____

_____

pampas

_____

_____

gauchos

_____

_____

Terms To Review

Use each of the following terms that you studied earlier in a sentence that reflects the term's meaning.

major
(Chapter 2, Section 3)

_____

_____

Water Systems *(pages 197–198)*

Questioning

As you read, write two questions about the main ideas presented in the text. After you have finished reading, write the answers to these questions.

1. _____

_____

_____

2. _____

_____

_____

Terms To Know

Define or describe the following key terms from this lesson.

hydroelectric power ▷ _____

_____

estuary ▷ _____

_____

Academic Vocabulary

Circle the letter of the word that has the closest meaning to the underlined word from this lesson.

transport ▷ **1.** The river systems of Latin America are used to <u>transport</u> people and goods throughout the region.

    a. explore   **b.** protect   **c.** move

volume ▷ **2.** The Amazon River carries ten times the <u>volume</u> of water that the Mississippi River carries.

    a. amount   **b.** influence   **c.** crowd

Natural Resources (page 198)

Skimming

Read the title and quickly look over the section to get a general idea of the section's content. Then write a sentence or two explaining what the section is about.

_____

_____

_____

Academic Vocabulary

Define these two academic vocabulary words from this lesson.

distributed
_____
_____

challenge
_____
_____

Section Wrap-up

Now that you have read the section, write the answers to the questions that were included in Setting a Purpose for Reading *at the beginning of the lesson.*

How do geographers divide the large region known as Latin America?

_____

_____

What factors have shaped Latin America's landforms?

_____

_____

How has the Latin American landscape influenced patterns of human settlement?

_____

_____

What natural resources make Latin America an economically important region?

_____

_____

_____

Chapter 8, Section 2
Climate and Vegetation

(Pages 199–203)

Reason To Read

Setting a Purpose for Reading Think about these questions as you read:
- Which climate regions are represented in Latin America?
- How do Latin America's location and landforms affect climates even within particular regions?
- How are the natural vegetation and agriculture of Latin America influenced by climatic factors?

Main Idea

As you read pages 199–203 in your textbook, complete this graphic organizer by filling in the characteristics of the three vertical climate zones.

Climate Zone	Characteristics
tierra caliente	

Climate and Vegetation Regions *(pages 199–202)*

Outlining   *Complete this outline as you read:*

I. Tropical Regions

    A. _____

    B. _____

II. The Rain Forest

    A. _____

    B. _____

III. Tropical Savanna

    A. _____

    B. _____

IV. The Humid Subtropics

    A. _____

    B. _____

V. Desert and Steppe Areas

    A. _____

    B. _____

Terms To Know   *Define or describe the following key term from this lesson.*

canopy _____

Places To Locate

Fill in the blank with the correct place from this lesson from the list below.

Argentina

Colombia

Amazon Basin

Atacama Desert

Venezuela

Uruguay

1. The _____ covers about one-third of South America and is the world's wettest tropical plain.

2. Two South American countries that have a tropical savanna climate and vast areas of llanos are _____ and _____ .

3. Two South American countries that have humid subtropical climates and pampas are _____ and _____ .

4. The _____ in Chile is a region so arid that in some places no rainfall has ever been recorded.

Academic Vocabulary

Define the following vocabulary words from this lesson.

dominate

_____

_____

_____

survey

_____

_____

Terms To Review

Write the letter of the correct definition in Group B next to the correct term that you studied earlier in Group A. One definition will not be used.

Group A

_____ **1.** prevailing winds
(Chapter 3, Section 2)

_____ **2.** rain shadow
(Chapter 3, Section 2)

Group B

a. dry area found on the leeward side of a mountain range

b. reason for thickness of rainforest vegetation

c. breeze in a region that blows in fairly constant directional pattern

Elevation and Climate (pages 202–203)

Monitoring Comprehension

As you read the lesson, write down questions you have about what you read. When you have finished reading the lesson, answer your questions.

_____

_____

_____

_____

Terms To Know

Fill in the blank with the correct term from this lesson from the list below.

Equator	*tierra fria*	*tierra caliente*
Sea Level	*tierra templada*	

1. The Spanish term for "hot land" is _____ .

2. The Spanish term for "temperate land" is _____ .

3. The Spanish term for "cold land" is _____ .

Terms To Review

Use each of these terms that you studied earlier in a sentence that reflects the term's meaning.

climate
(Chapter 3, Section 1)

_____

_____

occur
(Chapter 2, Section 2)

_____

_____

Academic Vocabulary

Define these two academic vocabulary words from this lesson.

affect

_____

_____

 **Key Points**

 **Notes**

annual

_____

_____

 **Section Wrap-up**

Now that you have read the section, write the answers to the questions that were included in Setting a Purpose for Reading *at the beginning of the lesson.*

Which climate regions are represented in Latin America?

_____

_____

_____

How do Latin America's location and landforms affect climates even within particular regions?

_____

_____

_____

How are the natural vegetation and agriculture of Latin America influenced by climatic factors?

_____

_____

_____

Chapter 9, Section 1
Population Patterns

(Pages 211–217)

Reason To Read

Setting a Purpose for Reading Think about these questions as you read:
- What ethnic groups make up the population of Latin America?
- How have geography and economics influenced the distribution of Latin America's population?
- How has migration affected the Latin American culture region?
- In what ways does Latin America's cultural diversity present both benefits and challenges for its people?

Main Idea

As you read pages 211–217 in your textbook, complete this graphic organizer by filling in the reasons Latin Americans migrate to the United States.

Reasons for Migrating North

Human Characteristics *(pages 211–213)*

Drawing Conclusions

As you read, write three details about the human characteristics of Latin America. Then write a general statement based on these details.

Details

_____

_____

_____

_____

General Statement

_____

_____

Terms To Know

Fill in the blank with the correct term from this lesson from the list below. Some terms will be used more than once.

indigenous

1. Each Latin American country has its own _____ , or form of a language unique to a particular place or group.

2. Native Americans today are descendants of the region's first inhabitants, so they are an _____ group.

dialect

3. Latin American forms of _____ blend indigenous, European, African, and Asian languages.

patois

4. The many forms of Latin American _____ are examples of _____ .

Academic Vocabulary

Define the following academic vocabulary words from this lesson.

temporary

_____

_____

communities

_____

_____

 **Key Points**

 **Notes**

Terms To Review

Use each of the following terms that you studied earlier in a sentence that reflects the term's meaning from this lesson.

cultures
(Chapter 1, Section 2)

_____

_____

ethnic
(Chapter 6, Section 3)

_____

_____

Where Latin Americans Live (pages 213–215)

Synthesizing

As you read, think about the main ideas of the section. Ask yourself, What effect does the high growth rate of Latin America have on the population density of the region?

_____

_____

Academic Vocabulary

Circle the letter of the word or phrase that has the closest meaning to the underlined word from this lesson.

1. Most <u>estimates</u> suggest that Latin America's population will grow to about 800 million by the year 2050.

   a. events   **b.** authorities   **c.** calculations

Terms To Review

Use the following terms that you studied earlier in a sentence that reflects the term's meaning from this lesson.

diverse
(Chapter 6, Section 1)

_____

_____

pampas
(Chapter 8, Section 1)

_____

_____

 **Key Points**

 **Notes**

Migration (pages 215–216)

Previewing

Preview the section to get an idea of what is ahead. First, skim the section. Then write a sentence or two describing what you think you will be learning. After you have finished reading, revise your statement as necessary.

_____

_____

_____

Terms To Know

Define or describe the following key term from this lesson.

> **urbanization**

_____

_____

Academic Vocabulary

Circle the letter of the word or phrase that has the closest meaning to the underlined words from this lesson.

1. People who migrate from Latin America often have a <u>conflict</u> because they want better job opportunities, but they do not want to leave their families and their homeland.

   a. go-between   **b.** struggle   **c.** choice

2. Many Latin American immigrants go through a process to enter the United States <u>legally</u>.

   a. lawfully   **b.** safely   **c.** quickly

Growth of Cities (pages 216–217)

Predicting

Read the title and main headings of the lesson. Write a statement predicting what the lesson is about and what will be included in the text. As you read, adjust or change your prediction if it does not match what you learn.

_____

_____

_____

Terms To Know

Define or describe the following key terms from this lesson.

megacities

_____

_____

primate city

_____

_____

Academic Vocabulary

Circle the letter of the word or phrase that has the closest meaning to the underlined words from this lesson.

1. The rapid growth of Mexico City makes it difficult to supply <u>adequate</u> housing for all the people.

   a. suitable   **b.** modern   **c.** costly

2. Countries in Latin America ship resources to countries <u>overseas</u>.

   a. across the rivers   **b.** across the ocean   **c.** across the continent

 *Now that you have read the section, write the answers to the questions that were included in* Setting a Purpose for Reading *at the beginning of the lesson.*

What ethnic groups make up the population of Latin America?

_____

_____

How have geography and economics influenced the distribution of Latin America's population?

_____

_____

How has migration affected the Latin American culture region?

_____

_____

In what ways does Latin America's cultural diversity present both benefits and challenges for its people?

_____

_____

116

Chapter 9, Section 1

Chapter 9, Section 2
History and Government
(Pages 220-225)

Reason To Read

Setting a Purpose for Reading Think about these questions as you read:
- What contributions have Latin America's Native American empires made to the region's cultural development?
- How has colonial rule influenced Latin America's political and social structures?
- How did most Latin American countries make the transition from colonialism to democracy?
- What political and social factors continue to challenge the Latin American culture region?

Main Idea

As you read pages 220-225 in your textbook, complete this graphic organizer by filling in characteristics of three Native American empires which existed in what is now Latin America.

Native American Empires
Maya
Aztec
Inca

Native American Empires (pages 220–222)

Scanning

Scan the lesson before you begin to read. As you glance quickly over the lines of text, look for key words or phrases that will tell you what the text will cover. Write the key words or phrases. Use the key words and phrases to write a statement explaining the lesson content. Revise your statement when you are finished reading the lesson.

> Key words or phrases

_____
_____
_____

> What the lesson is about

_____
_____
_____

Terms To Know

Write the letter of the correct definition from this lesson in Group B next to the correct term in group A. One definition will not be used.

Group A

_____ **1.** glyphs

_____ **2.** *chinampas*

_____ **3.** quipu

Group B

a. popular tourist attractions

b. a series of knotted cords of various colors and lengths

c. picture writings carved in stone

d. floating "islands" made from large rafts

Academic Vocabulary

Define the following academic vocabulary words from this lesson.

> established

_____
_____

> accurate

_____
_____

Empires to Nations (pages 222–224)

Sequencing

As you read, number the following events in the history of Latin America in the correct order.

_____ Spanish conquistador, Hernán Cortés, defeated the Aztec in Mexico.

_____ Mexico became independent.

_____ Haiti won independence from France.

_____ Cuba won independence from Spain.

_____ Spanish conquistador, Francisco Pizarro, destroyed the Inca Empire in Peru.

Terms To Know

Define or describe the following key terms from this lesson.

conquistador ⟩ _____
_____
_____

viceroys ⟩ _____
_____

Academic Vocabulary

Circle the letter of the word or phrase that has the closest meaning to the underlined word from this lesson.

1. The most important <u>unifying</u> institution in the Spanish and Portuguese colonies was the Roman Catholic Church.

    a. uniting   **b.** understanding   **c.** exciting

2. Native Americans in Latin America <u>converted</u> to Christianity.

    a. escaped   **b.** changed beliefs   **c.** kept going

Terms To Review

Use the following term that you studied earlier in a sentence that reflects the term's meaning from this lesson.

export
(Chapter 4, Section 4) ⟩ _____
_____

Era of Dictatorships (page 224)

Monitoring Comprehension

As you read the lesson, write down questions you have about what you read. When you have finished reading the lesson, answer your questions.

_____

_____

_____

_____

Terms To Know

Define or describe the following key terms from this lesson.

> caudillo

_____

_____

Movements for Change (pages 224–225)

Evaluating

When you evaluate something you read, you make a judgment or form an opinion. After you read this section, decide whether you agree or disagree with the quote by Rigoberta Menchú and tell why.

_____

_____

_____

Academic Vocabulary

Define the following academic vocabulary words from this lesson.

> issued

_____

_____

> military

_____

_____

 *Now that you have read the section, write the answers to the questions that were included in* Setting a Purpose for Reading *at the beginning of the lesson.*

What contributions have Latin America's Native American empires made to the region's cultural development?

_____

_____

How has colonial rule influenced Latin America's political and social structures?

_____

_____

How did most Latin American countries make the transition from colonialism to democracy?

_____

_____

What political and social factors continue to challenge the Latin American culture region?

_____

_____

Chapter 9, Section 3
Cultures and Lifestyles

(Pages 226–231)

Reason To Read

Setting a Purpose for Reading Think about these questions as you read:
- What role does religion play in Latin American culture?
- How have Latin Americans used the arts to express their history, their social struggles, and their cultural diversity?
- How is Latin America's cultural diversity reflected in family life, leisure activities, and public celebrations?

Main Idea

As you read pages 226–231 in your textbook, complete this graphic organizer by filling in a brief description of each aspect of Latin American culture.

Cultural Aspect	Description
Religion	
Arts	
Everyday Life	

Religion *(pages 226–228)*

Summarizing

As you read, complete the following sentences. Doing so will help you summarize the section.

1. During the _____ era, most Latin Americans became Christians.

2. Most Christians in the Spanish colonies and Brazil became

_____.

3. British and Dutch settlers brought _____ Christianity to Latin America.

4. Many Latin Americans blend their beliefs and practices into a

_____ faith.

Terms To Know

Define or describe the following key term from this lesson.

syncretism

_____

_____

Academic Vocabulary

Define the following academic vocabulary word from this lesson.

accompanied

_____

_____

Terms To Review

Use each of the following terms that you studied earlier in a sentence that reflects the term's meaning from this lesson.

role
(Chapter 7, Section 1)

_____

_____

traditions
(Chapter 4, Section 3)

_____

_____

The Arts of Latin America (pages 228–229)

Outlining

Complete this outline as you read.

The Arts of Latin America

I. Traditional Arts

   A. _____

   B. _____

II. Modern Arts

   A. _____

   B. _____

Terms To Know

Define or describe the following key terms from this lesson.

murals

_____

_____

mosaics

_____

_____

Academic Vocabulary

Define the following academic vocabulary words from this lesson.

symbols

_____

_____

abstract

_____

_____

Everyday Life (pages 229–231)

Interpreting

Think about the sense of loyalty to family that most Latin Americans have. Why do Latin Americans place so much emphasis on family life?

_____

_____

Terms To Know

Write the letter of the correct definition from this lesson in Group B next to the correct term in group A. One definition will not be used.

Group A

_____ **1.** extended family

_____ **2.** malnutrition

_____ **3.** fútbol

_____ **4.** jai alai

Group B

a. a group that includes grandparents, aunts, uncles, and cousins

b. a fast-paced game much like handball

c. soccer

d. festival

e. a serious condition caused by lack of proper food

Academic Vocabulary

Fill in the blank with the correct term from this lesson from the list below. One term will not be used.

status   traces   explicit

1. Latin American society shows _____ , or small hints, of machismo, a Spanish and Portuguese tradition of male supremacy.

2. Latin Americans place a heavy emphasis on social

_____ , or rank.

Terms To Review

Use the following term that you studied earlier in a sentence that reflects the term's meaning from this lesson.

literacy rate
(Chapter 6, Section 3)

_____

_____

Section Wrap-up

Now that you have read the section, write the answers to the questions that were included in Setting a Purpose for Reading *at the beginning of the lesson.*

What role does religion play in Latin American culture?

_____

_____

How have Latin Americans used the arts to express their history, their social struggles, and their cultural diversity?

_____

_____

How is Latin America's cultural diversity reflected in family life, leisure activities, and public celebrations?

_____

_____

Chapter 10, Section 1
Living in Latin America

(Pages 237–241)

Reason To Read

Setting a Purpose for Reading Think about these questions as you read:
- What is the basis of the economies of many Latin American countries?
- What are the advantages and disadvantages of the North American Free Trade Agreement (NAFTA) for Mexico?
- What are the causes and consequences of Latin America's economically dependent status?
- How has the region's physical geography affected transportation and communications?

Main Idea

As you read pages 237–241 in your textbook, complete this graphic organizer by listing factors that limit industrial growth.

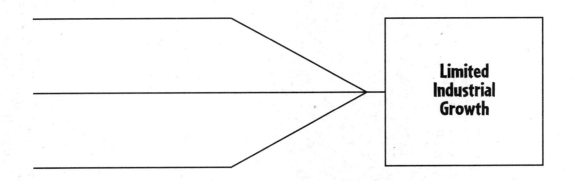

Agriculture (pages 237–238)

**Analyzing**

As you read this section, think about its organization and main ideas. Then write a sentence explaining the organization and list three main ideas.

Organization >

_____
_____
_____

Main Ideas >

_____
_____
_____

Terms To Know

Write the letter of the correct definition from Group B next to the correct term from this lesson in Group A. One definition will not be used.

Group A

_____ **1.** export

_____ **2.** campesinos

_____ **3.** _latifundia_

_____ **4.** _minifundia_

_____ **5.** cash crops

Group B

a. wealthy landowners

b. food or other goods grown in large quantities to be sold or traded

c. sell goods to other countries

d. large agricultural estates owned by wealthy families or corporations

e. rural farmers and workers

f. small plots of land intensively farmed by rural farmers to feed their families

Academic Vocabulary

Define the following academic vocabulary word from this lesson.

enable >

_____

 **Key Points**

Industry *(pages 238–240)*

Responding

As you read this section, think about what attracts your attention as you read. Write down facts you find interesting or surprising in the lesson.

_____

_____

_____

_____

_____

Terms To Know

Define or describe the following key terms from this lesson.

developing countries

_____

_____

service industries

_____

_____

maquiladoras

_____

_____

Terms To Review

Use the following term that you studied earlier in a sentence that reflects the term's meaning.

natural resources
(Chapter 4, Section 4)

_____

_____

Trade and Interdependence (page 240)

Clarifying

As you read this lesson, write down terms or concepts you find confusing. Then go back and reread the lesson to clarify the confusing parts. Write an explanation of the terms and concepts. If you are still confused, write your questions down and ask your teacher to clarify.

_____

_____

_____

Terms To Know

Define or describe the following key term from this lesson.

North American Free Trade Agreement (NAFTA) >

_____

_____

Academic Vocabulary

Define the following academic vocabulary words from this lesson.

obtain >

_____

_____

implemented >

_____

Transportation (pages 240–241)

Visualizing

Visualize the information described is this lesson to help you understand and remember what you have read. First, read the lesson. Next, ask yourself, What would this look like? Finally, write a description of the pictures you visualized on the lines below.

_____

_____

_____

Academic Vocabulary

Define the following academic vocabulary word from this lesson.

projects >

_____
_____

Communications *(page 241)*

Questioning

As you read, write two questions about the main ideas presented in the text. After you have finished reading, write the answers to these questions.

1. _____
_____
_____
_____

2. _____
_____
_____
_____

Academic Vocabulary

Define the following academic vocabulary word from this lesson.

technology >

_____
_____

Terms To Review

Use the following term that you studied earlier in a sentence that reflects the term's meaning from this lesson.

networks >
(Chapter 1, Section 1)

_____
_____

Section Wrap-up

Now that you have read the section, write the answers to the questions that were included in Setting a Purpose for Reading *at the beginning of the lesson.*

What is the basis of the economies of many Latin American countries?

_____

_____

_____

What are the advantages an disadvantages of the North American Free Trade Agreement (NAFTA) for Mexico?

_____

_____

_____

What are the causes and consequences of Latin America's economically dependent status?

_____

_____

_____

How has the region's physical geography affected transportation and communications?

_____

_____

_____

Chapter 10, Section 2
People and Their Environment

(Pages 242–247)

Reason To Read

Setting a Purpose for Reading Think about these questions as you read:
- How has development affected Latin America's forest resources?
- How are Latin American governments working to balance forest conservation with human and economic development?
- What challenges are posed by the growth of Latin America's urban population?
- What regional and international issues continue to pose challenges for Latin American countries?

Main Idea

As you read pages 242–247 in your textbook, complete this outline.

I. Managing Rain Forests

    A. _____

    B. _____

II. Urban Environments

    A. _____

    B. _____

III. Regional and International Issues

    A. _____

    B. _____

    C. _____

    D. _____

Managing Rain Forests (pages 242–244)

(pages 242–244)

Evaluating

When you evaluate something you read, you make a judgment or form an opinion. After you read this lesson, decide whether you agree or disagree with the quote by George E. Stuart, and explain your opinion.

_____

_____

_____

Terms To Know

Fill in the blank with the correct term from this lesson from the list below. One term will be used twice.

sustainable development

1. Many Latin American countries are working toward _____

_____ because this technological and economic growth does not deplete the human and natural resources of an area.

deforestation

2. _____ is the planting of young trees or the seeds of trees on the land that has been stripped.

slash-and-burn

3. The _____ method puts nutrients into the soil, but this benefit is only temporary as nutrients are washed away within one or two years.

reforestation

4. Many new settlers to the rain forest use a method of clearing the land

called _____, in which they cut down all plants, strip the trees of bark, and when the trees have dried out, set them on fire.

5. _____ is the clearing or destruction of forests.

Academic Vocabulary

Define the following academic vocabulary words from this lesson.

strategies

_____

_____

debated

> **Terms To Review**

Use the following term that you studied earlier in a sentence that reflects the term's meaning from this lesson.

indigenous
(Chapter 9, Section 1)

Urban Environments *(pages 244–245)*

> **Previewing**

Preview the lesson to get an idea of what's ahead. First, skim the lesson. Then write a sentence or two explaining what you think you will be learning. After you have finished reading, revise your statements as necessary.

> **Terms To Know**

Define or describe the following key term from this lesson.

shantytowns

> **Places To Locate**

Explain why the following place from this lesson is important.

São Paulo

 Key Points

 Notes

Academic Vocabulary

Define the following academic vocabulary word from this lesson.

exceed >

_____

_____

Terms To Review

Use each of the following terms that you studied earlier in a sentence that reflects the term's meaning from this lesson.

megacities
(Chapter 9, Section 1) >

_____

_____

_____

malnutrition
(Chapter 9, Section 3) >

_____

_____

Regional and International Issues (pages 245–247)

Interpreting

Think about the effects of industrial pollution in Latin America. How can Latin America overcome the challenges of industrial pollution?

_____

_____

Places To Locate

Explain why the following place from this lesson is important.

El Salvador >

_____

_____

Academic Vocabulary

Define the following academic vocabulary words from this lesson.

incentives >

_____

_____

 Key Points

 Notes

ultimately

 Terms To Review

Use each of the following terms that you studied earlier in a sentence that reflects the term's meaning from this lesson.

conflicts
(Chapter 9, Section 1)

birthrates
(Chapter 4, Section 1)

Section Wrap-up

Now that you have read the section, write the answers to the questions that were included in **Setting a Purpose for Reading** *at the beginning of the lesson.*

How has development affected Latin America's forest resources?

How are Latin American governments working to balance forest conservation with human and economic development?

What challenges are posed by the growth of Latin America's urban population?

What regional and international issues continue to pose challenges for Latin American countries?

Chapter 11, Section 1
The Land
(Pages 271–276)

Reason To Read

Setting a Purpose for Reading Think about these questions as you read:
- Why is Europe sometimes called a "peninsula of peninsulas"?
- What are some of the numerous islands surrounding the continent of Europe?
- Why are rivers vital to Europe's economy?
- What are some of Europe's most important natural resources?

Main Idea

As you read pages 271–276 in your textbook, complete this graphic organizer by filling in the natural resources found in Europe.

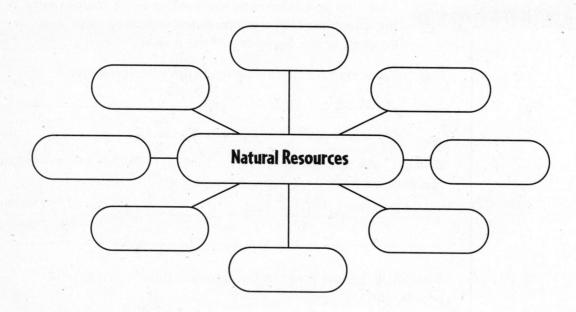

Seas, Peninsulas, and Islands (pages 271–274)

Visualizing

Visualize the information described is this lesson to help you understand and remember what you have read. First, read the lesson. Next, ask yourself, **What would this look like?** *Finally, write a description of the pictures you visualized on the lines below.*

_____

_____

_____

Terms To Know

Write the letter of the correct definition in Group B next to the correct term from this lesson in Group A. One definition will not be used.

Group A	Group B
_____ **1.** dikes	**a.** large banks of earth and stone
_____ **2.** polders	**b.** an irregular coastline
_____ **3.** glaciation	**c.** a process in which large bodies of ice form and spread
_____ **4.** fjords	**d.** long, narrow, steep-sided inlets
	e. reclaimed lands

Academic Vocabulary

Circle the letter of the word or phrase that has the closest meaning to the underlined word from this lesson.

distinct

1. Although Europe and Asia share a common landmass called Eurasia, Europe is a <u>distinct</u> region.

   a. well-defined   **b.** larger   **c.** neighboring

remove

2. Pumps <u>remove</u> water from polders.

   a. put back   **b.** break up   **c.** take away

Key Points

Notes

Terms To Review

Use the following term that you studied earlier in a sentence that reflects the term's meaning from this lesson.

glaciers
(Chapter 2, Section 2)

_____

_____

Mountains and Plains *(pages 274–275)*

Scanning

Scan the lesson before you begin to read. As you glance quickly over the lines of text, look for key words or phrases that will tell you what the text will cover. Write the key words or phrases. Then use the key words and phrases to write a statement explaining the lesson content. Revise your statement when you are finished reading the lesson.

Key words or phrases

_____

_____

_____

What the lesson is about

_____

_____

_____

_____

Terms To Review

Use the following term that you studied earlier in a sentence that reflects the term's meaning from this lesson.

ranges
(Chapter 1, Section 2)

_____

_____

Water Systems *(page 275–276)*

Inferring

As you read the lesson, look for clues in the descriptions and events that might help you draw a conclusion to the question: How does Europe's network of rivers and canals contribute to industrial development in the region?

_____

_____

_____

Academic Vocabulary

Define the following academic vocabulary words from this lesson.

enhanced

_____

_____

_____

thereby

_____

_____

Terms To Review

Use the following term that you studied earlier in a sentence that reflects the term's meaning from this lesson.

tributary
(Chapter 5, Section 1)

_____

_____

Natural Resources *(page 276)*

Skimming

Read the title and quickly look over the lesson to get a general idea of the lesson's content. Then write a sentence or two explaining what the lesson is about.

_____

_____

_____

Academic Vocabulary

Define the following academic vocabulary words from this lesson.

nuclear

_____
_____

invested

_____
_____

Terms To Review

Use the following term that you studied earlier in a sentence that reflects the term's meaning from this lesson.

hydroelectric power
(Chapter 8, Section 1)

_____
_____

Section Wrap-up

Now that you have read the section, write the answers to the questions that were included in Setting a Purpose for Reading *at the beginning of the lesson.*

Why is Europe sometimes called a "peninsula of peninsulas"?

_____
_____

What are some of the numerous islands surrounding the continent of Europe?

_____
_____

Why are rivers vital to Europe's economy?

_____
_____

What are some of Europe's most important natural resources?

_____
_____

Chapter 11, Section 2
Climate and Vegetation
(Pages 277–281)

Reason To Read

Setting a Purpose for Reading Think about these questions as you read:
- What are the climate regions in Europe?
- What physical features influence Europe's climates?
- Why are most of Europe's original forests gone?

Main Idea

As you read pages 277–281 in your textbook, complete this graphic organizer by listing the types of climate regions found in the mid-latitudes.

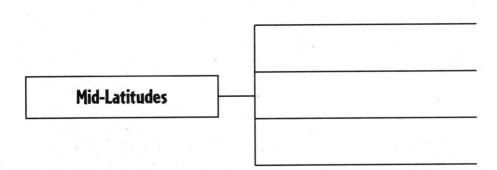

Water and Land (pages 277–278)

Monitoring Comprehension

As you read the lesson, write down questions you have about what you read. When you have finished reading the lesson, answer your questions.

_____

_____

_____

_____

Terms To Review

Use the following terms that you studied earlier in a sentence that reflects the term's meaning from the lesson.

factors
(Chapter 4, Section 2)

_____

_____

region
(Chapter 1, Section 1)

_____

_____

Western Europe (pages 278–280)

Predicting

Read the title and main headings of the lesson. Write a statement predicting what the lesson is about and what will be included in the text. As you read, adjust or change your prediction if it does not match what you learn.

_____

_____

_____

Terms To Know

Fill in the blank with the correct term from this lesson from the list below. One term will be used twice.

timberline

1. Dry winds called _____ blow down from the mountains into valleys and plains, causing sudden changes in weather.

2. The elevation above which trees cannot grow is called the

foehns

_____ .

3. _____ are destructive masses of ice, snow, and rock that slide down mountainsides.

avalanches

4. Coniferous trees grow in the cooler Alpine mountain region up to the

_____ , beyond which it is too cold for them to survive.

Terms To Review

Use the following terms that you studied earlier in a sentence that reflects the term's meaning from this lesson.

prevailing winds
(Chapter 3, Section 2)

_____

_____

reforestation
(Chapter 10, Section 2)

_____

_____

Southern Europe (pages 280–281)

Synthesizing

As you read, think about the main ideas of the lesson. Ask yourself, if I wanted to vacation in southern Europe, when would be the best time to go? Explain why.

_____

_____

_____

Key Points

Notes

Terms To Know

Fill in the blank with the correct term from this lesson from the list below. One term will be used twice.

> **mistral**

> **siroccos**

> **chaparral**

1. High, dry winds from North Africa, called _____, bring hot temperatures to southern Europe.

2. A strong north wind from the Alps is called the _____.

3. _____, such as cork oak and olive trees, grow in southern Europewhere the climate is hot and dry in the summer.

4. The _____ sometimes sends gusts of bitterly cold air into southern France.

Terms To Review

Use the following term that you studied earlier in a sentence that reflects the term's meaning.

> **precipitation**
> (Chapter 2, Section 3)

_____

_____

Eastern and Northern Europe *(page 281)*

Drawing Conclusions

As you read the lesson, write three details about the climate of eastern and northern Europe. Then write a conclusion based on these details.

> **Details**

_____

_____

_____

> **Conclusion**

_____

_____

Terms To Know

Define or describe the following key term from this lesson.

permafrost

_____

_____

Section Wrap-up

Now that you have read the section, write the answers to the questions that were included in Setting a Purpose for Reading *at the beginning of the lesson.*

What are the climate regions in Europe?

_____

_____

What physical features influence Europe's climates?

_____

_____

Why are most of Europe's original forests gone?

_____

_____

Chapter 12, Section 1
Population Patterns

(Pages 287–291)

Reason To Read

Setting a Purpose for Reading Think about these questions as you read:
- How does the physical geography of Europe influence its population density and distribution?
- What effects has industrialization and urbanization had on Europe's people?
- How have recent patterns of migration influenced the region's culture?

Main Idea

As you read pages 287–291 in your textbook, complete this graphic organizer by listing the features that contribute to an average to higher-than-average population density.

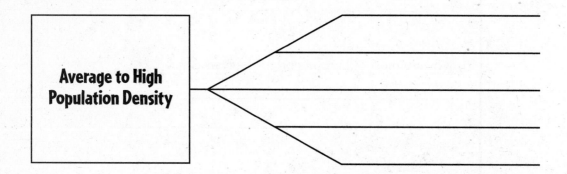

Average to High
Population Density

Ethnic Diversity *(pages 287–289)*

Drawing Conclusions

As you read, find details in the lesson that support the following conclusion: **Europeans think of themselves as Europeans as well as members of ethnic and/or national groups.**

> **Details**

_____

_____

_____

Terms To Know

Fill in the blank with the correct term from this lesson from the list below. One term will be used twice.

> **ethnic groups**

> **ethnic cleansing**

> **refugees**

1. People who flee to a foreign country for safety are _____ .

2. Europe has more than 160 _____ , or groups of people who share ancestry, language, customs, and often religion.

3. During the fighting between the republic of Bosnia-Herzegovina and Serb-ruled Kosovo, Serb leaders followed a policy of _____ , in which they expelled or killed rival ethnic groups in these areas.

4. Two Belgian _____ are the Flemings and the Walloons.

Academic Vocabulary

Define the following academic vocabulary words from this lesson.

> **external**

_____

_____

> **internal**

_____

_____

Terms To Review

Use the following terms that you studied earlier in a sentence that reflects the term's meaning from this lesson.

trace
(Chapter 9, Section 3)

_____

_____

_____

republic
(Chapter 6, Section 2)

_____

_____

Population Characteristics *(page 289)*

Synthesizing

As you read, think about the main ideas of the lesson. Ask yourself, what human activity has caused the greatest population density in areas of Europe?

_____

_____

_____

_____

Terms To Review

Use the following terms that you studied earlier in a sentence that reflects the term's meaning from this lesson.

population density
(Chapter 4, Section 1)

_____

_____

distributed
(Chapter 8, Section 1)

_____

_____

Urbanization *(pages 289–291)*

Evaluating

As you read, form an opinion about the information in this lesson. Does the quote by Erla Zwingle support the following statement? Explain why or why not.

Naples, Italy, is a city that reflects the coming together of past and present in modern Europe.

_____

_____

_____

_____

Terms To Know

Define or describe the following key term from this lesson.

urbanization

_____

_____

Academic Vocabulary

Define the following academic vocabulary words from this lesson.

expert

_____

_____

predict

_____

_____

Terms To Review

Use the following term that you studied earlier in a sentence that reflects the term's meaning from this lesson.

available
(Chapter 2, Section 3)

_____

_____

   Now that you have read the section, write the answers to
the questions that were included in **Setting a Purpose for**
Reading *at the beginning of the lesson.*

How does the physical geography of Europe influence its population
density and distribution?

_____

_____

_____

_____

What effects has industrialization and urbanization had on Europe's
people?

_____

_____

_____

_____

How have recent patterns of migration influenced the region's culture?

_____

_____

_____

_____

Chapter 12, Section 2
History and Government

(Pages 294–300)

Reason To Read

Setting a Purpose for Reading Think about these questions as you read:

- What contributions did early Europeans make to world culture?
- In what ways has Europe's geography shaped its history?
- What were the effects of world wars and economic and political revolutions in Europe?

Main Idea

As you read pages 294–300 in your textbook, complete this graphic organizer by recording key events in Europe's history for each date.

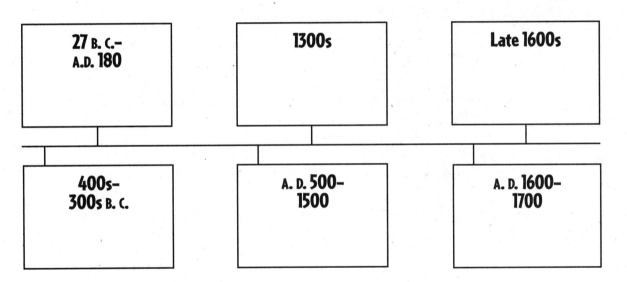

| 27 B. C.–A.D. 180 | 1300s | Late 1600s |

| 400s–300s B. C. | A. D. 500–1500 | A. D. 1600–1700 |

The Rise of Europe (pages 294–296)

Scanning

Scan the lesson before you begin to read. As you glance quickly over the lines of text, look for key words or phrases that will tell you what the text will cover. Write the key words or phrases. Then use the key words and phrases to write a statement explaining the lesson content. Revise your statement when you are finished reading the lesson.

Key words or phrases ⟩ _____
_____
_____
_____

What the lesson is about ⟩ _____
_____
_____
_____

Terms To Know

Write the letter of the correct definition in Group B next to the correct term from this lesson in group A. One definition will not be used.

Group A

_____ **1.** city-states

_____ **2.** Middle Ages

_____ **3.** feudalism

Group B

a. a system in which monarchs or lords gave land to nobles in return for pledges of loyalty

b. the period between ancient and modern times

c. an artificial channel for carrying water

d. in ancient Greece, an independent community consisting of a city and the surrounding lands

 # Key Points

 # Notes

Academic Vocabulary

Define the following academic vocabulary words from this lesson.

> **contacts**

_____
_____
_____

> **foundations**

_____
_____
_____

Terms To Review

Use the following terms that you studied earlier in a sentence that reflects the term's meaning from this lesson.

> **democracy**
> (Chapter 4, Section 3)

_____
_____

> **republic**
> (Chapter 6, Section 2)

_____
_____

Expansion of Europe (pages 296–297)

Summarizing

As you read, complete the following sentences. This will help you summarize the lesson.

1. Beginning in the 1000s, western European armies fought the

_____ to win _____ ,
the birthplace of Christianity, from Muslim rule.

2. Beginning in the 1300s, the _____ brought about great advances in European civilization.

3. The _____ , which began in the 1400s, led to the

beginnings of _____ and a lessening of the power of the Roman Catholic Church.

Terms To Know

Define or describe the following key terms from this lesson.

Crusades

_____

_____

_____

Renaissance

_____

_____

_____

Reformation

_____

_____

_____

Academic Vocabulary

Define the following academic vocabulary words from this lesson.

series

_____

_____

_____

routes

_____

_____

_____

A Changing Europe (pages 297–300)

Responding

As you read this lesson, think about what attracts your attention as you read. Write down facts you find interesting or surprising in the lesson.

_____

_____

_____

_____

_____

Terms To Know

Write the letter of the correct definition in Group B next to the correct term from this lesson in group A. One definition will not be used.

Group A

_____ **1.** Enlightenment

_____ **2.** industrial capitalism

_____ **3.** communism

_____ **4.** reparation

_____ **5.** Holocaust

_____ **6.** Cold War

_____ **7.** European Union (EU)

Group B

a. the mass killing of more than 6 million European Jews and others by Germany's Nazi leaders

b. a philosophy for a society based on economic equality in which the workers control factories and production

c. payment for damages

d. a power struggle between the communist world, and the noncommunist world

e. an East German barrier built to stop the movement of people to West Germany

f. organization dedicated to a united Europe in which goods, services, and workers move freely

g. an economic system in which business leaders use profits to expand their companies

h. movement by educated Europeans emphasizing the importance of reason and questioning long-standing values

Terms To Review

Use the following terms that you studied earlier in a sentence that reflects the term's meaning from this lesson.

revolutions
(Chapter 3, Section 1)

_____

_____

market economies
(Chapter 4, Section 3)

_____

_____

Section Wrap-up

Now that you have read the section, write the answers to the questions that were included in Setting a Purpose for Reading *at the beginning of the lesson.*

What contributions did early Europeans make to world culture?

_____

_____

_____

_____

In what ways has Europe's geography shaped its history?

_____

_____

_____

_____

What were the effects of world wars and economic and political revolutions in Europe?

_____

_____

_____

_____

Chapter 12, Section 3
Cultures and Lifestyles

(Pages 301–307)

Reason To Read

Setting a Purpose for Reading Think about these questions as you read:
- How has religion influenced the cultural development of Europe?
- Why has European art and culture been so influential throughout the world?
- How do European governments meet the educational and health care needs of their peoples?

Main Idea

As you read pages 301–307 in your textbook, complete this graphic organizer by filling in the major headings of the section.

I. Expressions of Culture

    A. _____

    B. _____

    C. _____

II. Quality of Life

    A. _____

    B. _____

III. Lifestyles

    A. _____

    B. _____

    C. _____

 Notes

Expressions of Culture (pages 301–305)

Clarifying

As you read this lesson, write down terms or concepts you find confusing. Then go back and reread the lesson to clarify the confusing sections. Write an explanation of the terms and concepts. If you are still confused, and ask your teacher to clarify.

_____

_____

_____

_____

Terms To Know

Fill in the blank with the correct term from this lesson from the list below.

dialect   romanticism   language family

impressionists   realism   welfare state

Good Friday Peace Agreement

1. A _____ is a group of related languages that developed from an earlier language.

2. A _____ is a local form of a language.

3. A group of painters called _____ paint out-doors to capture immediate experiences of the natural world.

4. An artistic style that focuses on accurately depicting the details of everyday life is called _____ .

5. An artistic style that focuses on emotions, stirring historical events, and the exotic is called _____ .

6. The _____ was an understanding between Protestant and Roman Catholic communities to share political power.

Places To Locate

Explain why the following places from this lesson are important.

Switzerland

_____

_____

_____

Northern Ireland

_____

_____

Academic Vocabulary

Define the following academic vocabulary words from this lesson.

reveals

_____

_____

_____

focused

_____

_____

Terms To Review

Use the following term that you studied earlier in a sentence that reflects the term's meaning from this lesson.

abstract
(Chapter 9, Section 3)

_____

_____

Quality of Life (pages 305–306)

Determining the Main Idea

As you read, write the main idea of the lesson. Review your statement when you have finished reading and revise as necessary.

_____

_____

_____

Terms To Know

Define or describe the following key term from this lesson.

welfare states

_____

_____

Academic Vocabulary

Circle the letter of the word or phrase that has the closest meaning to the underlined word from this lesson.

1. The gap in the quality of life among various parts of Europe <u>poses</u> an obstacle to full European unity.

   a. suggests   **b.** presents   **c.** shows

2. Throughout Europe, the number of years of <u>required</u> schooling varies from country to country.

   a. called for   **b.** optional   **c.** invited

Terms To Review

Use the following terms that you studied earlier in a sentence that reflects the term's meaning from this lesson.

literacy rates
(Chapter 6, Section 3)

_____

_____

internal
(Chapter 12, Section 1)

_____

_____

Lifestyles (pages 306–307)

Connecting

As you read, compare the lifestyles of people in Europe with the lifestyles of people in the United States. Summarize your thoughts in a paragraph. Be sure to include ways that European lifestyles are similar to and different from American lifestyles.

_____

_____

_____

 **Key Points**

 **Notes**

Academic Vocabulary

Define the following academic vocabulary word from this lesson.

revolves

_____

_____

Terms To Review

Use the following terms that you studied earlier in a sentence that reflects the term's meaning from this lesson.

solstice
(Chapter 3, Section 1)

_____

_____

urbanization
(Chapter 12, Section 1)

_____

_____

Section Wrap-up

Now that you have read the section, write the answers to the questions that were included in **Setting a Purpose for Reading** *at the beginning of the lesson.*

How has religion influenced the cultural development of Europe?

_____

_____

Why has European art and culture been so influential throughout the world?

_____

_____

How do European governments meet the educational and health care needs of their peoples?

_____

_____

Chapter 13, Section 1
Living in Europe
(Pages 313–319)

Reason To Read

Setting a Purpose for Reading Think about these questions as you read:
- What economic systems are found in Europe?
- Why are economic changes taking place in Europe?
- How do transportation and communications systems link European countries to each other and to the rest of the world?

Main Idea

As you read pages 313–319 in your textbook, complete this graphic organizer by filling in the goals of the European Union.

Goals of EU

Changing Economies (pages 313–316)

Reviewing

As you read the lesson, fill in the cause-and-effect chart below. Review your chart when you have finished reading and revise as needed.

Cause	Effect

Terms To Know

Define or describe the following key terms from this lesson.

European Union (EU) > _____

Maastricht Treaty > _____

Academic Vocabulary

Define the following academic vocabulary words from this lesson.

policy > _____

welfare > _____

Terms To Review

Use the following terms that you studied earlier in a sentence that reflects the term's meaning from this lesson.

market economies
(Chapter 4, Section 3) > _____

command economies
(Chapter 4, Section 3) > _____

Industry *(pages 316–317)*

Responding

As you read this lesson, think about what attracts your attention as you read. Write down facts you find interesting or surprising in the lesson.

_____

_____

_____

_____

Terms To Know

Define or describe the following key terms from this lesson.

heavy industry

_____

_____

light industry

_____

_____

Academic Vocabulary

Circle the letter of the word or phrase that has the closest meaning to the underlined word from this lesson.

1. High-technology industries are a growing <u>sector</u> of western Europe's economy.

   a. guide   **b.** style   **c.** division

Terms To Review

Use the following term that you studied earlier in a sentence that reflects the term's meaning.

linked
(Chapter 5, Section 1)

_____

_____

Agriculture (pages 317–318)

Analyzing

As you read this lesson, think about the organization and main ideas of the text. Then write a sentence explaining the organization and list the main ideas.

Organization

_____

_____

Main Ideas

_____

_____

Terms To Know

Write the letter of the correct definition in Group B next to the correct term from this lesson in Group A. One definition will not be used.

Group A	Group B
____ **1.** mixed farming	**a.** organizations in which farmers share in growing and selling products
____ **2.** farm cooperatives	**b.** food or other goods grown in large quantities to be sold or traded
____ **3.** collective farms	**c.** foods with genes altered to make them grow bigger, faster, and resistant to pests
____ **4.** state farms	**d.** using natural substances instead of fertilizers and chemicals to increase crop yields
____ **5.** genetically modified foods	**e.** raising several kinds of crops and livestock on the same farm
____ **6.** organic farming	**f.** a state-owned farm managed by government officials in which workers receive wages but no share of products or profits
	g. state-owned farm on which farmers receive wages plus a share of products and profits

Academic Vocabulary

Define the following academic vocabulary word from this lesson.

consumer

_____

_____

Transportation and Communications (pages 318–319)

Outlining

Complete this outline as you read.

I. Railways and Highways

   A. _____

   B. _____

II. _____

   A. _____

   B. _____

III. _____

   A. _____

   B. _____

Terms To Review

Use the following term that you studied earlier in a sentence that reflects the term's meaning from this lesson.

tributaries
(Chapter 5, Section 1)

_____

_____

Section Wrap-up

Now that you have read the section, write the answers to the questions that were included in Setting a Purpose for Reading *at the beginning of the lesson.*

What economic systems are found in Europe?

_____

Why are economic changes taking place in Europe?

_____

_____

How do transportation and communications systems link European countries to each other and to the rest of the world?

_____

_____

Chapter 13, Section 2
People and Their Environment
(Pages 320–325)

Reason To Read

Setting a Purpose for Reading Think about these questions as you read:

- How have industry and farming practices affected Europe's environment?
- What steps are being taken to protect Europe's environment?
- What successes have Europeans had in recent decades in reversing the effects of pollution?

Main Idea

As you read pages 320–325 in your textbook, complete this graphic organizer by listing some of the reasons why Eastern Europe has pollution problems.

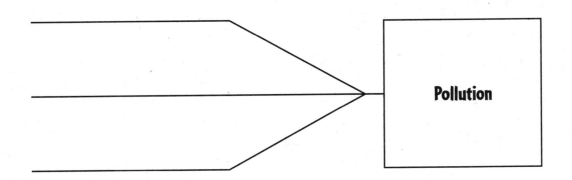

Pollution

Humans and the Environment (pages 320–321)

Reviewing

Preview the lesson to get an idea of what is ahead. First, skim the lesson. Then write a sentence or two explaining what you think the text is about. After you have finished reading, revise your statements as necessary.

_____

_____

_____

Terms To Know

Define or describe the following key term from this lesson.

> **dry farming**

_____

_____

Academic Vocabulary

Define the following academic vocabulary word from this lesson.

> **indicates**

_____

_____

Terms To Review

Use the following terms that you studied earlier in a sentence that reflects the term's meaning from this lesson.

> **climate**
> (Chapter 3, Section 1)

_____

_____

Pollution *(pages 321–323)*

Predicting

Read the title and main headings of the lesson. Write a statement predicting what the lesson is about and what will be included in the text. As you read, adjust or change your prediction if it does not match what you learn.

_____

_____

_____

Terms To Know

Fill in the blank with the correct term from this lesson from the list below. One term will be used twice.

acid deposition	greenhouse effect	environmentalists
acid rain	global warming	meltwater

1. Some scientists believe that _____ is causing the earth's average temperature to rise.

2. People concerned with the effects of pollution on the earth are called

_____ .

3. The _____ is created by carbon dioxide and other gases that trap the sun's heat near the earth's surface.

4. _____ is acid pollution that mixes with moisture in the air and falls to the ground.

5. The result of melting snow and ice is _____ .

6. Precipitation, called _____ , carries large amounts of dissolved acids, which damage buildings, forests, and crops, and kills wildlife.

7. Many _____ are studying the effects of increased carbon dioxide in the earth's atmosphere.

Key Points

Notes

Academic Vocabulary

Define the following academic vocabulary word from this lesson.

compounds >

_____

_____

Terms To Review

Use the following terms that you studied earlier in a sentence that reflects the term's meaning from this lesson.

atmosphere
(Chapter 2, Section 1) >

_____

_____

issue
(Chapter 9, Section 2) >

_____

_____

Reducing Pollution (pages 323–325)

Interpreting

Think about the impact of pollution in Europe. What are some of Europe's major challenges as its countries work to improve the environment?

_____

_____

Places To Locate

Explain why the following place from this lesson is important.

Carpathian
Mountains >

_____

_____

Academic Vocabulary

Define the following academic vocabulary words from this lesson.

sites >

_____

_____

Key Points

Notes

concept

_____

_____

Terms To Review

Use the following terms that you studied earlier in a sentence that reflects the term's meaning from this lesson.

legal
(Chapter 9, Section 1)

_____

_____

European Union (EU)
(Chapter 13, Section 1)

_____

_____

Section Wrap-up

Now that you have read the section, write the answers to the questions that were included in Setting a Purpose for Reading *at the beginning of the lesson.*

How have industry and farming practices affected Europe's environment?

_____

_____

What steps are being taken to protect Europe's environment?

_____

_____

What successes have Europeans had in recent decades in reversing the effects of pollution?

_____

_____

Chapter 14, Section 1
The Land

(Pages 345–350)

Reason To Read

Setting a Purpose for Reading Think about these questions as you read:
- How large is the land area of Russia?
- How does Russia's interconnected plains and mountain ranges shape settlement in the country?
- What are Russia's natural resources?

Main Idea

As you read pages 345–350 in your textbook, complete this outline.

The Land

I. A Vast and Varied Land

   A. _____

   B. _____

   C. _____

II. Rivers

   A. _____

   B. _____

III. Natural Resources

   A. _____

   B. _____

   C. _____

A Vast and Varied Land (pages 345–348)

Visualizing

Visualize the information described in this lesson to help you understand and remember what you have read. First, read the lesson. Next, ask yourself, What would this look like? *Finally, write a description of the pictures you visualized on the lines below.*

_____

_____

_____

Terms To Know

Define or describe the following key terms from this lesson.

chernozem

_____

_____

_____

hydroelectric power

_____

_____

Places To Locate

Write the letter of the correct location in Group B next to the correct place from this lesson in Group A. One location will not be used.

Group A

____ **1.** Caucasus Mountains

____ **2.** Central Siberian Plateau

____ **3.** North European Plain

____ **4.** West Siberian Plain

Group B

a. tableland area in Siberia

b. mountain range in southwestern Russia that lies between the Black and the Caspian Seas

c. area of flat land that stretches from the Arctic Ocean to the grasslands of Central Asia

d. area of flat land that sweeps across western and central Europe into Russia

e. mountain range that is the traditional boundary between European Russia and Asian Russia

Academic Vocabulary

Circle the letter of the word or phrase that has the closest meaning to the underlined word from this lesson.

1. The Caspian Sea is a saltwater lake that <u>occupies</u> a deep depression.

   a. blocks off   **b.** goes around   **c.** fills up

2. Water in the Caspian Sea evaporates and leaves behind salts that <u>accumulate</u> and make the water saltier.

   a. give way   **b.** add up   **c.** take away

Terms To Review

Use the following terms that you studied earlier in a sentence that reflects the term's meaning from this lesson.

series
(Chapter 12, Section 2)

_____

_____

location
(Chapter 1, Section 1)

_____

_____

Rivers *(pages 348–349)*

Determining the Main Idea

As you read, write the main idea of the lesson. Review your statement when you have finished reading and revise as needed.

_____

_____

Terms To Know

Define or describe the following key term from this lesson.

hydroelectric power

_____

_____

Notes

Academic Vocabulary

Define the following academic word from this lesson.

generated >

_____

_____

Terms To Know

Use the following terms that you studied earlier in a sentence that reflects the term's meaning from this lesson.

portion
(Chapter 3, Section 2) >

_____

_____

meltwaters
(Chapter 13, Section 2) >

_____

_____

Natural Resources *(page 349–350)*

Inferring

As you read the lesson, look for clues in the descriptions and events that might help you draw a conclusion to the question: **Why do 75 percent of Russians live west of the Ural Mountains?**

_____

_____

_____

_____

Terms To Know

Define or describe the following key term from this lesson.

permafrost >

_____

_____

Academic Vocabulary

Define the following academic vocabulary words from this lesson.

utilize

_____
_____
_____

underlies

_____
_____
_____

Terms To Review

Use the following terms that you studied earlier in a sentence that reflects the term's meaning from this lesson.

physical geography
(Chapter 1, Section 2)

_____
_____
_____

layer
(Chapter 3, Section 3)

_____
_____
_____

Section Wrap-up

Now that you have read the section, write the answers to the questions that were included in **Setting a Purpose for Reading** _at the beginning of the lesson._

How large is the land area of Russia?

_____
_____

How does Russia's interconnected plains and mountain ranges shape settlement in the country?

_____
_____

What are Russia's natural resources?

_____
_____

Chapter 14, Section 2
Climate and Vegetation
(Pages 351–355)

Reason To Read

Setting a Purpose for Reading Think about these questions as you read:
- What are Russia's major climates?
- What are the seasons like in Russia?
- How does climate affect the way Russians live?
- What types of natural vegetation are found in each of Russia's climate regions?

Main Idea

As you read pages 351–355 in your textbook, complete this graphic organizer by describing the climates and vegetation of Russia.

Region	Description
Tundra	
Taiga	
Steppe	

Russia's Climates and Vegetation *(pages 351–352)*

Monitoring Comprehension

As you read the lesson, write down questions you have about what you read. When you have finished reading the lesson, answer your questions.

_____

_____

_____

_____

Places To Locate

Explain why the following place from this lesson is important.

Siberia

_____

_____

Terms To Review

Use the following term that you studied earlier in a sentence that reflects the term's meaning from the lesson.

occur
(Chapter 2, Section 2)

_____

_____

High Latitude Climates *(pages 352–353)*

Scanning

Scan the lesson before you begin to read. As you glance quickly over the lines of text, look for key words or phrases that will tell you what the text will cover. Write the key words or phrases. Then use the key words and phrases to write a statement explaining the lesson content. Revise your statement when you are finished reading the lesson.

Key words or phrases

_____

_____

_____

Key Points

Notes

> **What the lesson is about**

_____

_____

_____

Terms To Know

Fill in the blank with the correct term from this lesson from the list below. One term will be used twice.

tundra   taiga

1. Russia's far north is the _____, or a vast, treeless plain.

2. The vegetation in the subarctic region of Russia is _____, which is a type of forest.

3. The Russian _____ is the world's largest coniferous forest.

Places To Locate

Explain why the following place from the lesson is important.

> **Arctic Circle**

_____

_____

Academic Vocabulary

Define the following academic vocabulary words from this lesson.

> **adjustments**

_____

_____

> **construct**

_____

_____

Terms To Review

Use the following terms that you studied earlier in a sentence that reflects the term's meaning from this lesson.

isolated
(Chapter 8, Section 1)

_____
_____
_____

dominates
(Chapter 8, Section 2)

_____
_____

Mid-Latitude Climates (pages 354–355)

Synthesizing

As you read, think about the main ideas of the lesson. What are the similarities and differences between the humid continental and steppe climate regions?

_____
_____
_____
_____

Terms To Know

Define or describe the following key term from this lesson.

steppe

_____
_____

Academic Vocabulary

Define the following academic vocabulary words from this lesson.

encountered

_____
_____

region

_____

Terms To Review

Use the following terms that you studied earlier in a sentence that reflects the term's meaning.

coniferous
(Chapter 3, Section 3)

_____

_____

mixed forests
(Chapter 3, Section 3)

_____

_____

Section Wrap-up

Now that you have read the section, write the answers to the questions that were included in Setting a Purpose for Reading *at the beginning of the lesson.*

What are Russia's major climates?

_____

_____

What are the seasons like in Russia?

_____

_____

How does climate affect the way Russians live?

_____

_____

What types of natural vegetation are found in each of Russia's climate regions?

_____

_____

Chapter 15, Section 1
Population Patterns

(Pages 363–366)

Reason To Read

Setting a Purpose for Reading Think about these questions as you read:
• What ethnic groups make up Russia's population?
• Why is Russia's population unevenly distributed?
• How does the climate east of the Ural Mountains affect the population distribution in this region?

Main Idea

As you read pages 363–366 in your textbook, complete this graphic organizer by filling in the major ethnic groups in Russia.

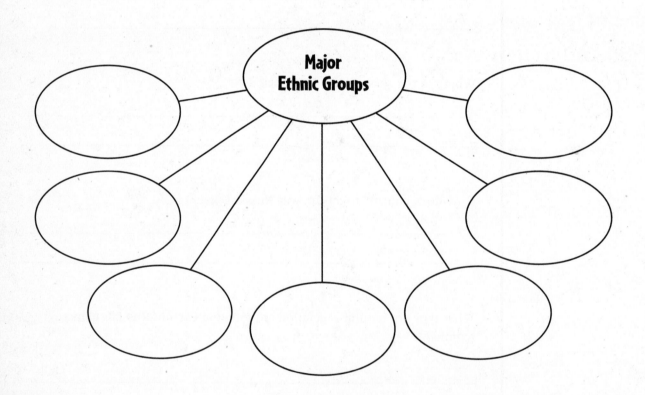

Russia's Ethnic Diversity (pages 363–365)

Evaluating

As you read, form an opinion about the information in this lesson. Does the quote by Mike Edwards support the following statement? **The Tartar population in Russia is growing rapidly.** *Explain your answer.*

_____

_____

_____

_____

Terms To Know

Define or describe the following key terms from this lesson.

ethnic group
_____

_____

nationalities
_____

_____

sovereignty
_____

_____

Academic Vocabulary

Define the following academic vocabulary words from this lesson.

ratio
_____

_____

granted
_____

_____

Terms To Review

Use the following terms that you studied earlier in a sentence that reflects the term's meaning from this lesson.

concentrated
(Chapter 4, Section 1)
_____

_____

erode
(Chapter 5, Section 2)

_____

_____

Population Density and Distribution (pages 365–366)

Skimming

Read the title and quickly look over the lesson to get a general idea of the lesson's content. Then write a sentence or two explaining the lesson's content.

_____

_____

Places To Locate

Explain why the following places from this lesson are important.

Ural Mountains

_____

_____

Moscow

_____

Academic Vocabulary

Define the following academic vocabulary words from this lesson.

appreciate

_____

_____

trend

_____

_____

Terms To Review

Use the following terms that you studied earlier in a sentence that reflects the term's meaning from this lesson.

population density
(Chapter 4, Section 1)

_____

_____

environment
(Chapter 1, Section 1)

Now that you have read the section, write the answers to the questions that were included in Setting a Purpose for Reading *at the beginning of the lesson.*

What ethnic groups make up Russia's population?

_____

_____

_____

_____

Why is Russia's population unevenly distributed?

_____

_____

_____

_____

How does the climate east of the Ural Mountains affect the population distribution in this region?

_____

_____

_____

_____

Chapter 15, Section 2
History and Government

(Pages 367–373)

Reason To Read

Setting a Purpose for Reading Think about these questions as you read:
• Who were the ancestors of the ethnic Russians?
• Why did the rule of the czars end in revolution?
• What were the causes of the Soviet Union's collapse?
• Why does Russia face an uncertain future?

Main Idea

As you read pages 367–373 in your textbook, complete this graphic organizer by recording major events in Russia's history for each date.

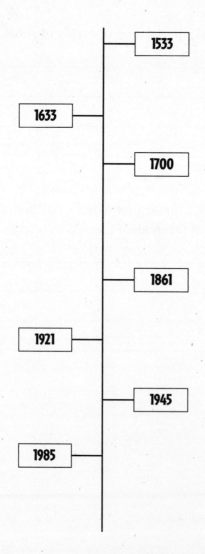

	1533
1633	
	1700
	1861
1921	
	1945
1985	

Early Peoples and States (pages 367–369)

Summarizing

As you read, complete the following sentences. Doing so will help you summarize the lesson.

1. The Varangians settled among the Slavs living near the

   _____ and _____ Rivers.

2. In the early 1200s, _____ invaders from Central

   Asia conquered Kiev and many of the Slav territories.

3. By the late 1400s, the _____ drove the Mongols

   from Muscovy.

Terms To Know

Define or describe the following key terms from this lesson.

czar
_____
_____
_____

serfs
_____
_____

Romanov Czars (pages 369–370)

Questioning

As you read, write two questions about the main ideas presented in the text. After you have finished reading, write the answers to these questions.

1. _____

   _____

2. _____

   _____

Academic Vocabulary

Circle the letter of the word or phrase that has the closest meaning to the underlined word from this lesson.

1. By the late 1700s, the Russian nobility had made French the <u>primary</u> language of the Russian empire.

   a. main   **b.** earliest   **c.** only

The Russian Revolution (page 370)

Evaluating

When you evaluate something you read, you make a judgment or form an opinion. After you read this lesson, decide if you agree or disagree with the ideas of Karl Marx and explain why.

_____

_____

_____

_____

Terms To Know

Define or describe the following key terms from this lesson.

> **Russification**

_____

_____

> **socialism**

_____

_____

The Soviet Era (pages 370–371)

Synthesizing

As you read, think about the main ideas of the lesson. Ask yourself, how were the governments during czarist rule and the Soviet era similar and different?

_____

_____

_____

_____

Terms To Know

Write the letter of the correct definition in Group B next to the correct term from this lesson in group A. One definition will not be used.

Group A

_____ **1.** Bolsheviks

_____ **2.** communism

_____ **3.** satellites

_____ **4.** Cold War

Group B

a. a civil war

b. a philosophy that called for the violent overthrow of government and the creation of a new society led by workers

c. countries controlled by the Soviet Union

d. a power struggle between the communist world and the noncommunist world

e. a revolutionary group led by Vladimir Ilyich Lenin

Academic Vocabulary

Circle the letter of the word or phrase that has the closest meaning to the underlined word from this lesson.

The Bolsheviks tried to <u>maintain</u> power in Russia by treating their opponents harshly.

a. end   **b.** hold onto   **c.** increase

The Soviet Breakup *(pages 371–372)*

Responding

As you read this lesson, think about what attracts your attention as you read. Write down facts you find interesting or surprising in the lesson.

_____

_____

_____

_____

Terms To Know

Define or describe the following key terms from this lesson.

perestroika

_____

_____

_____

glasnost

_____

_____

Academic Vocabulary

Circle the letter of the word or phrase that has the closest meaning to the underlined word from this lesson.

After the formation of the Commonwealth of Independent States, the Soviet Union <u>ceased</u> to exist.

a. established   **b.** claimed   **c.** stopped

A New Russia *(pages 372–373)*

Predicting

Read the title and main headings of the lesson. Write a statement predicting what the lesson is about and what will be included in the text. As you read, adjust or change your prediction if it does not match what you learn.

_____

_____

Now that you have read the section, write the answers to the questions that were included in Setting a Purpose for Reading *at the beginning of the lesson.*

Who were the ancestors of the ethnic Russians?

_____

_____

Why did the rule of the czars end in revolution?

_____

_____

What were the causes of the Soviet Union's collapse?

_____

_____

Why does Russia face an uncertain future?

_____

_____

Chapter 15, Section 3
Cultures and Lifestyles

(Pages 376–381)

Reason To Read

Setting a Purpose for Reading Think about these questions as you read:
- How has the role of religion changed in post-Soviet Russia?
- How are education and health care in Russia adjusting to the fall of communism?
- What role do art, music, and literature have in Russia's cultural heritage?

Main Idea

As you read pages 376–381 in your textbook, complete this graphic organizer by filling in information about religions in Russia.

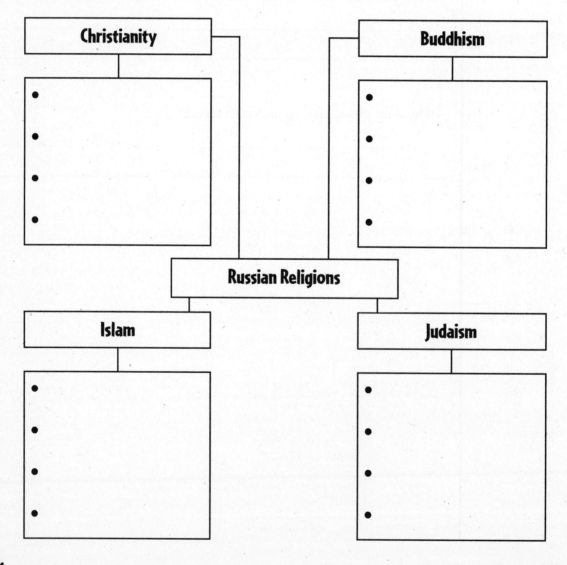

Religion in Russia *(pages 376–378)*

Clarifying

As you read this lesson, write down terms or concepts you find confusing. Then go back and reread the lesson to clarify the confusing parts. Write an explanation of the terms and concepts. If you are still confused, write your questions down and ask your teacher to clarify.

_____

_____

_____

Terms To Know

Fill in the blank with the correct term from this lesson from the list below. One term will be used twice.

atheism

1. _____ were attacks on Jews carried out by government troops or officials in czarist Russia.

2. _____ is the belief that there is no God.

patriarch

3. The head of the Eastern Orthodox Church is called a

_____ .

icon

4. An _____ is a religious image or symbol.

pogrom

5. The Soviet government discouraged religion and encouraged

_____ .

Academic Vocabulary

Define the following academic vocabulary words from this lesson.

restrictions

_____

_____

images

_____

_____

Terms To Review

Use the following term that you studied earlier in a sentence that reflects the term's meaning from this lesson.

restore
(Chapter 5, Section 2)

_____

_____

Education (pages 378–379)

Determining the Main Idea

As you read, write the main idea of the lesson. Review your statement when you have finished reading and revise as needed.

_____

_____

_____

Terms To Know

Define or describe the following key term from this lesson.

intelligentsia

_____

_____

Academic Vocabulary

Circle the letter of the word or phrase that has the closest meaning to the underlined word from this lesson.

Today, schools in Russia emphasize a more <u>objective</u> and less authoritative approach to learning.

a. goal **b.** strict **c.** impartial

Health Care (page 379)

Previewing

Preview the lesson to get an idea of what is ahead. First, skim the lesson. Then write a sentence or two explaining what you think you will be learning. After you have finished reading, revise your statements as necessary.

_____

_____

_____

_____

The Arts (pages 379–381)

Scanning

Scan the lesson before you begin to read. As you glance quickly over the lines of text, look for key words or phrases that will tell you what the text will cover. Write the key words or phrases. Then use the key words and phrases to write a statement explaining the lesson content. Revise your statement when you are finished reading the lesson.

Key words or phrases >

_____

_____

_____

What the lesson is about >

_____

_____

_____

Academic Vocabulary

Define the following academic vocabulary word from this lesson.

guidelines >

_____

_____

Life and Leisure *(page 381)*

Connecting

As you read, compare the lifestyles of people in Russia with the lifestyles of people in the United States. Summarize your thoughts in a paragraph. Be sure to include ways that Russian lifestyles are similar to and different from American lifestyles.

_____

_____

_____

_____

Terms To Review

Use the following term that you studied earlier in a sentence that reflects the term's meaning from this lesson.

despite
(Chapter 8, Section 1)

_____

_____

Section Wrap-up

Now that you have read the section, write the answers to the questions that were included in Setting a Purpose for Reading *at the beginning of the lesson.*

How has the role of religion changed in post-Soviet Russia?

_____

_____

How are education and health care in Russia adjusting to the fall of communism?

_____

_____

What role do art, music, and literature have in Russia's cultural heritage?

_____

_____

Chapter 16, Section 1
Living in Russia
(Pages 387–399)

Reason To Read

Setting a Purpose for Reading Think about these questions as you read:
- How has Russia made the transition to a market economy?
- How have agriculture, industry, transportation, and communications in Russia changed since the breakup of the Soviet Union?
- What is Russia's relationship to the global community?

Main Idea

As you read pages 387–393 in your textbook, complete this graphic organizer by listing changes in Russia's economic system and the effect of each.

Economic System	Effect
Command Economy	
Market Economy	

Changing Economies (pages 387–390)

Reviewing

As you read the lesson, look for cause-and-effect relationships and fill in the cause-and-effect chart below. Review your chart when you have finished reading and revise as needed.

Cause	Effect

Terms To Know

Write the letter of the correct definition in Group B next the correct term from this lesson in Group A. One definition will not be used.

Group A

_____ **1.** command economy

_____ **2.** consumer goods

_____ **3.** black market

_____ **4.** market economy

_____ **5.** privatization

Group B

a. goods needed for everyday life

b. an economy in which a central authority makes key economic decisions

c. any illegal place where scarce or illegal goods are sold, usually at high prices

d. goods such as tanks and other military hardware, machinery, and electric generators

e. a change to private ownership of state-owned companies and industries

f. an economy in which businesses are privately owned

Terms To Review

Use the following term that you studied earlier in a sentence that reflects the term's meaning from this lesson.

invest
(Chapter 11, Section 1)

_____

_____

Agriculture and Industry *(pages 390–391)*

Monitoring Comprehension

As you read the lesson, write down questions you have about what you read. When you have finished reading the lesson, answer your questions.

_____

_____

_____

_____

_____

Terms To Know

Define or describe the following key terms from this lesson.

kolkhoz

_____

_____

_____

sovkhoz

_____

_____

Academic Vocabulary

Circle the letter of the word or phrase that has the closest meaning to the underlined word from this lesson.

1. For many years, Russia's state-owned aerospace industry and its military-industrial system were its economic and <u>technical</u> focus.

   a. mechanical   **b.** suggested   **c.** brief

2. Russia's <u>domestic</u> oil provides its other industries with needed energy at a reasonable cost.

   a. from the homeland   **b.** exported   **c.** imported

Terms To Review

Use the following terms that you studied earlier in a sentence that reflects the term's meaning.

sectors
(Chapter 13, Section 1)

_____

_____

Transportation and Communications (pages 391–393)

Interpreting

Think about transprtation in Russia. Then ask yourself: **How has climate effected transportation in Russia and how are Russians overcoming their transportation obstacles?** *Explain your answer.*

_____

_____

_____

_____

Academic Vocabulary

Define the following academic vocabulary word from this lesson.

complex

_____

_____

Terms To Review

Use the following terms that you studied earlier in a sentence that reflects the term's meaning from this lesson.

developed countries
(Chapter 4, Section 4)

_____

_____

steppes
(Chapter 14, Section 2)

_____

_____

Chapter 16, Section 1

Global Interdependence *(page 393)*

Inferring

As you read the lesson, look for clues in the descriptions and events that might help you draw a conclusion to the following question: **What can you infer about Russia's goals, based on changes in Russia's trade and international relations since the Soviet breakup?**

_____

_____

_____

_____

Terms To Review

Use the following term that you studied earlier in a sentence that reflects the term's meaning from this lesson.

republics
(Chapter 6, Section 2)

_____

_____

Section Wrap-up

Now that you have read the section, write the answers to the questions that were included in Setting a Purpose for Reading *at the beginning of the lesson.*

How has Russia made the transition to a market economy?

_____

_____

How have agriculture, industry, transportation, and communications in Russia changed since the breakup of the Soviet Union?

_____

_____

What is Russia's relationship to the global community?

_____

_____

Chapter 16, Section 2
People and Their Environment

(Pages 396–399)

Reason To Read

Setting a Purpose for Reading Think about these questions as you read:
- How does Russia manage its natural resources?
- How has pollution affected the lives of Russia's people?
- What are the environmental challenges in Russia's future?

Main Idea

As you read pages 396–399 in your textbook, complete this graphic organizer by describing the environmental issues and concerns for each location.

Location	Description
Kamchatka	
Lake Baikal	
Chernobyl	

Managing Resources (pages 396–397)

Determining the Main Idea

As you read, write the main idea of the lesson. Review your statement when you have finished reading and revise as needed.

_____

_____

Places To Locate

Explain why the following place from this lesson is important.

Kamchatka >

_____

_____

Academic Vocabulary

Define the following academic vocabulary words from this lesson.

target >

_____

_____

aware >

_____

_____

Terms To Review

Use the following terms that you studied earlier in a sentence that reflects the term's meaning from this lesson.

jobs
(Chapter 1, Section 2) >

_____

_____

challenge
(Chapter 8, Section 1) >

_____

_____

Pollution (pages 397–399)

Scanning

Scan the lesson before you begin to read. As you glance quickly over the lines of text, look for key words or phrases that will tell you what the text will cover. Write the key words or phrases. Then use the key words and phrases to write a statement explaining the lesson content. Revise your statement when you are finished reading the lesson.

Key words or phrases

_____
_____
_____
_____

What the lesson is about

_____
_____
_____
_____

Terms To Know

Define or describe the following key terms from this lesson.

radioactive material

_____
_____
_____

pesticides

_____
_____
_____

nuclear wastes

_____
_____
_____

Places To Locate

Explain why the following place from this lesson is important.

Lake Baikal

_____
_____

Academic Vocabulary

Define the following academic vocabulary word from this lesson.

exposed

_____

_____

Terms To Review

Use the following terms that you studied earlier in a sentence that reflects the term's meaning from this lesson.

prevailing winds
(Chapter 3, Section 2)

_____

_____

experts
(Chapter 12, Section 1)

_____

_____

Section Wrap-up

Now that you have read the section, write the answers to the questions that were included in Setting a Purpose for Reading at the beginning of the lesson.

How does Russia manage its natural resources?

_____

_____

How has pollution affected the lives of Russia's people?

_____

_____

What are the environmental challenges in Russia's future?

_____

_____

Chapter 17, Section 1
The Land
(Pages 421–426)

Reason To Read

Setting a Purpose for Reading Think about these questions as you read:
- What land and water features dominate the region?
- Why are the region's major rivers important to its people?
- Why is much of the world economically dependent on the region?

Main Idea

As you read pages 421–426 in your textbook, complete this graphic organizer by filling in a description of each body of water listed.

Body of Water	Description
Dead Sea	
Caspian Sea	
Aral Sea	

 Key Points

 Notes

Seas and Peninsulas (pages 421–422)

Clarifying

As you read this lesson, write down terms or concepts you find confusing. Then go back and reread the lesson to clarify the confusing parts. Write an explanation of the terms and concepts. If you are still confused, write your questions down and ask your teacher to clarify.

_____

_____

_____

Terms To Review

Use the following term that you studied earlier in a sentence that reflects the term's meaning from this lesson.

evaporation
(Chapter 2, Section 3)

_____

_____

Rivers (pages 422–424)

Inferring

As you read the lesson, look for clues in the descriptions and events that might help you draw a conclusion to the question: Why were the Nile Delta and the Tigris-Euphrates river valley "cradles of civilization"?

_____

_____

_____

_____

Terms To Know

Define or describe the following key terms from this lesson.

alluvial soil

_____

_____

wadis

_____

_____

Plains, Plateaus, and Mountains *(pages 424–425)*

Visualizing

Visualize the information described in this lesson to help you understand and remember what you have read. First, read the lesson. Next, ask yourself, What would this look like? Finally, write a description of the pictures you visualized on the lines below.

_____

_____

_____

_____

Terms To Know

Define or describe the following key term.

kums

_____

_____

Terms To Review

Use the following term that you studied earlier in a sentence that reflects the term's meaning from this lesson.

rain shadow
(Chapter 3, Section 2)

_____

_____

Earthquakes *(page 425)*

Responding

As you read this lesson, think about what attracts your attention as you read. Write down facts you find interesting or surprising in the lesson.

_____

_____

_____

_____

 **Key Points**

 **Notes**

Academic Vocabulary

Define the following academic vocabulary word from this lesson.

> **shift**

_____

_____

Terms To Review

Use the following term that you studied earlier in a sentence that reflects the term's meaning from this lesson.

> **region**
> (Chapter 14, Section 2)

_____

_____

Natural Resources *(page 426)*

Skimming

Read the title and quickly look over the lesson to get a general idea of the lesson's content. Then write a sentence or two explaining what the lesson covers.

_____

_____

Terms To Know

Define or describe the following key term.

> **phosphate**

_____

_____

Academic Vocabulary

Circle the letter of the word or phrase that has the closest meaning to the underlined word from this lesson.

1. When oil prices <u>fluctuate</u> on world markets, the region's economies suffer.

    a. remain constant   **b.** go up and down   **c.** stabilize

2. Oil, natural gas, and minerals bring <u>revenue</u> to the region.

    a. money, income   **b.** imported goods   **c.** ocean-going ships

Terms To Review

Use the following term that you studied earlier in a sentence that reflects the term's meaning from this lesson.

exports
(Chapter 10, Section 1)

_____

_____

Section Wrap-up

Now that you have read the section, write the answers to the questions that were included in Setting a Purpose for Reading *at the beginning of the lesson.*

What land and water features dominate the region?

_____

_____

Why are the region's major rivers important to its people?

_____

_____

Why is much of the world economically dependent on the region?

_____

_____

Chapter 17, Section 2
Climate and Vegetation

(Pages 427–431)

Reason To Read

Setting a Purpose for Reading Think about these questions as you read:
• How do the climates of North Africa, Southwest Asia, and Central Asia differ?
• How have the needs of a growing population affected the natural vegetation of the region?

Main Idea

As you read pages 427–431 in your textbook, complete this graphic organizer by filling in the three mid-latitude climate regions of North Africa, Southwest Asia, and Central Asia.

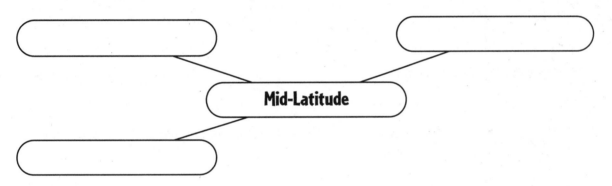

Mid-Latitude

Water: A Precious Resource (pages 427–430)

Monitoring Comprehension

As you read the lesson, write down questions you have about what you read. When you have finished reading the lesson, answer your questions.

_____
_____
_____
_____

Terms To Know

Define or describe the following key terms from this lesson.

oasis
_____
_____
_____

pastoralism
_____
_____

Places To Locate

Explain why the following places from this lesson are important.

Sahara
_____
_____
_____

Rub' al Khali
_____
_____

Garagum (Kara Kum)
_____
_____

Terms To Review

Use the following terms that you studied earlier in a sentence that reflects the term's meaning from the lesson.

evaporation
(Chapter 2, Section 3)

_____
_____
_____

steppe
(Chapter 14, Section 2)

_____
_____

Climatic Variations *(pages 430–431)*

Evaluating

As you read, form an opinion about the information in this lesson. Answer the following questions.

1. Is Thor Heyerdahl a reliable source? Explain your answer.

_____
_____

2. Does the quote by Thor Heyerdahl support the following statement? *Under the pressure of climate changes, grassy plains in the region turned into desert.* Explain your opinion.

_____
_____

Terms To Know

Define or describe the following key term from this lesson.

cereals

_____
_____

Terms To Review

Use the following term that you studied earlier in a sentence that reflects the term's meaning from this lesson.

pollution
(Chapter 4, Section 4)

_____

_____

Section Wrap-up

Now that you have read the section, write the answers to the questions that were included in Setting a Purpose for Reading *at the beginning of the lesson.*

How do the climates of North Africa, Southwest Asia, and Central Asia differ?

_____

_____

_____

_____

How have the needs of a growing population affected the natural vegetation of the region?

_____

_____

_____

_____

Chapter 18, Section 1
Population Patterns

(Pages 439–443)

Reason To Read

Setting a Purpose for Reading Think about these questions as you read:
- How have movement and interaction of people in the region led to ethnic diversity?
- How do the region's seas, rivers, and oases influence where people live?
- What effect does the growing migration into the cities have on the region?

Main Idea

As you read pages 439–443 in your textbook, complete this graphic organizer by naming the major ethnic groups of North Africa, Southwest Asia, and Central Asia.

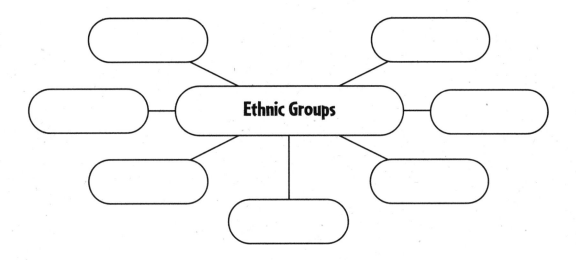

Many Peoples (pages 439–442)

Previewing

Preview the lesson to get an idea of what is ahead. First, skim the lesson. Then write a sentence or two explaining what you think you will be learning. After you have finished reading, revise your statements as necessary.

_____

_____

_____

Terms To Know

Define or describe the following key term from this lesson.

ethnic diversity ＞ _____

_____

Places To Locate

Explain why the following places from this lesson are important.

Turkey ＞ _____

_____

Afghanistan ＞ _____

_____

Armenia ＞ _____

_____

Georgia ＞ _____

_____

Kazakhstan ＞ _____

_____

Tajikistan ＞ _____

 **Key Points**

 **Notes**

Uzbekistan

_____

_____

Academic Vocabulary

Define the following academic vocabulary words from this lesson.

significant

_____

_____

_____

proportion

_____

_____

Terms To Review

Use the following terms that you studied earlier in a sentence that reflects the term's meaning from this lesson.

define
(Chapter 4, Section 3)

_____

_____

impact
(Chapter 7, Section 2)

_____

_____

Population and Resources *(pages 442–443)*

Synthesizing

As you read, think about the main ideas of the lesson. Ask yourself, **What might happen if Israel returns the Golan Heights to Syria? How would this affect life in Israel? In Syria?**

_____

_____

_____

_____

 Key Points

 Notes

Terms To Know

Define or describe the following key term from this lesson.

infrastructure

_____

_____

Places To Locate

Explain why the following place from this lesson is important.

Tehran

_____

_____

Terms To Review

Use the following terms that you studied earlier in a sentence that reflects the term's meaning from this lesson.

oases
(Chapter 3, Section 3)

_____

_____

overall
(Chapter 4, Section 1)

_____

_____

Section Wrap-up

Now that you have read the section, write the answers to the questions that were included in Setting a Purpose for Reading *at the beginning of the lesson.*

How have movement and interaction of people in the region led to ethnic diversity?

_____

_____

How do the region's seas, rivers, and oases influence where people live?

_____

_____

What effect does the growing migration into the cities have on the region?

_____

_____

Chapter 18, Section 2
History and Government

(Pages 446–452)

Reason To Read

Setting a Purpose for Reading Think about these questions as you read:
- What great civilizations arose in North Africa, Southwest Asia, and Central Asia?
- What three major world religions originated in the region?
- How did countries of the region gain independence in the modern era?

Main Idea

As you read pages 446–452 in your textbook, complete this graphic organizer by listing the achievements of each ancient civilization.

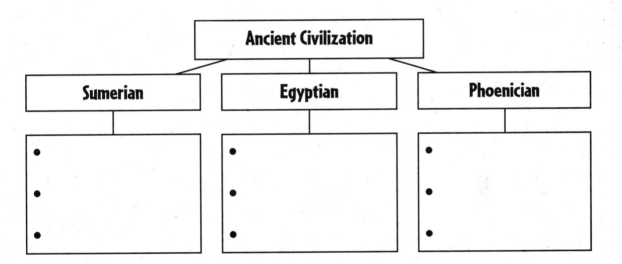

Prehistoric Peoples *(page 446)*

Drawing Conclusions

As you read the lesson, look for clues to help you answer the following question: **Why was the domestication of plants and animals so important for the early peoples in the region?**

> **Conclusion**

_____

_____

Terms To Know

Define or describe the following key term from this lesson.

> **domesticate**

_____

_____

Early Civilizations *(page 447)*

Questioning

As you read, write two questions about the main ideas presented in the text. After you have finished reading, write the answers to these questions.

1. _____

_____

_____

2. _____

_____

_____

Terms To Know

Fill in the blank with the correct term from this lesson from the list below. One term will be used twice.

> **cuneiform**

> **hieroglyphics**

> **culture hearths**

1. The Sumerians kept records by using wedge-shaped symbols called

_____.

2. The Mesopotamian civilization was one of the world's first

_____, or centers where ideas and traditions developed and spread outward.

3. The Egyptians developed a form of picture writing called

_____.

4. _____ flourished along the Nile and Tigris-Euphrates Rivers.

Academic Vocabulary

Circle the letter of the word or phrase that has the closest meaning to the underlined word from this lesson.

The Sumerians created a <u>code</u> of law to keep order.

a. nation   **b.** force   **c.** system

Empires and Trade (page 448)

Visualizing

Visualize the information described in this lesson to help you understand and remember what you have read. First, read the lesson. Next, ask yourself, **What would the Silk Road have looked like in 100 B.C.?** *Finally, write a description of the pictures you visualized on the lines below.*

_____

_____

_____

Terms To Know

Define or describe the following key term from this lesson.

> qanats

_____

_____

Three Major Religions (448–449)

Determining the Main Idea

As you read, write the main idea of the lesson. Review your statement when you have finished reading and revise as needed.

_____

_____

Terms To Know

Write the letter of the correct definition in Group B next to the correct term from this lesson in group A. One definition will not be used.

Group A

____ **1.** monotheism

____ **2.** prophets

____ **3.** mosque

Group B

a. a house of worship where Muslims pray

b. belief in one god

c. Islam's holy book

d. people who are believed to deliver messages from God

Academic Vocabulary

Define the academic vocabulary words from this lesson.

bond ➤ _____

_____

The Modern Era *(pages 449–452)*

Sequencing

As you read, number the following events in the correct sequential order.

____ Central Asian countries win their freedom with the breakup of the Soviet Union.

____ United States-led forces remove Iraq's military-based dictator, Saddam Hussein, from Iraq.

____ Israel is founded.

____ The Romans expel most Jews from Palestine.

____ The Wye River Agreement is signed.

 **Key Points**

 **Notes**

Terms To Know

Define or describe the following key terms from this lesson.

nationalism

_____
_____
_____

nationalized

_____
_____
_____

embargo

_____
_____
_____

Academic Vocabulary

Circle the letter of the word or phrase that has the closest meaning to the underlined word from this lesson.

A West Bank resident was <u>straightforward</u> when he said: "Israelis and Palestinians claim the right of return to the same land."

a. dishonest   **b.** difficult   **c.** direct

Section Wrap-up

Now that you have read the section, write the answers to the questions that were included in **Setting a Purpose for Reading** *at the beginning of the lesson.*

What great civilizations arose in North Africa, Southwest Asia, and Central Asia?

_____
_____

What three major world religions originated in the region?

_____
_____

How did countries of the region gain independence in the modern era?

_____
_____

Chapter 18, Section 3
Cultures and Lifestyles
(Pages 453–457)

Reason To Read

Setting a Purpose for Reading Think about these questions as you read:
- How have religion and language both unified and divided the peoples of North Africa, Southwest Asia, and Central Asia?
- What arts are popular in the region?
- What are some characteristics of everyday life in the region?

Main Idea

As you read pages 453–457 in your textbook, complete this graphic organizer by listing some of the region's famous literary works.

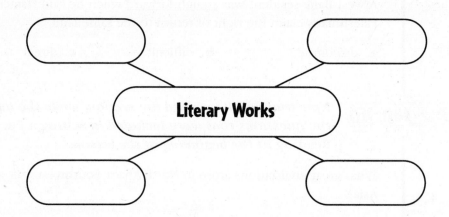

Religion (pages 453–454)

Scanning

Scan the lesson before you begin to read. As you glance quickly over the lines of text, look for key words or phrases that will tell you what the text will cover. Write the key words or phrases. Then use the key words and phrases to write a statement explaining the lesson content. Revise your statement when you have finished reading the lesson.

Key words or phrases

_____

_____

_____

What the lesson is about

_____

_____

_____

Terms To Review

Use the following term that you studied earlier in a sentence that reflects the term's meaning from this lesson.

community
(Chapter 9, Section 1)

_____

_____

Languages (page 454)

Determining the Main Idea

As you read, write the main idea of the lesson. Review your statement when you have finished reading and revise as needed.

_____

_____

_____

The Arts (pages 454–455)

Previewing

Preview the lesson to get an idea of what is ahead. First, skim the lesson. Then write a sentence or two explaining what you think you will be learning. After you have finished reading, revise your statements as necessary.

_____

_____

_____

Terms To Know

Define or describe the following key term from this lesson.

ziggurats

_____

_____

Terms To Review

Use the following term that you studied earlier in a sentence that reflects the term's meaning from this lesson.

designs
(Chapter 6, Section 1)

_____

_____

Everyday Life (pages 455–457)

Connecting

As you read, compare the lifestyles of people in Southwest Asia, North Africa, and Central Asia with the lifestyles of people in the United States. Summarize your thoughts in a paragraph. Be sure to include ways that the region's lifestyles are similar to and different from American lifestyles.

_____

_____

_____

_____

Terms To Know

Define or describe the following key terms from this lesson.

bedouins

_____
_____
_____

bazaar

_____
_____

Places To Locate

Explain why the following places from this lesson are important.

Qatar

_____
_____
_____

United Arab Emirates

_____
_____

Academic Vocabulary

Define the following academic vocabulary words from this lesson.

medical

_____
_____
_____

predominantly

_____
_____

Terms To Review

Use the following terms that you studied earlier in a sentence that reflects the term's meaning from this lesson.

contact
(Chapter 12, Section 2)

_____
_____

despite
(Chapter 8, Section 1)

_____
_____

Now that you have read the section, write the answers to the questions that were included in Setting a Purpose for Reading *at the beginning of the lesson.*

How have religion and language both unified and divided the peoples of North Africa, Southwest Asia, and Central Asia?

_____

_____

_____

_____

What arts are popular in the region?

_____

_____

_____

_____

What are some characteristics of everyday life in the region?

_____

_____

_____

_____

Chapter 19, Section 1
Living in North Africa, Southwest Asia, and Central Asia

(Pages 463–468)

Reason To Read

Setting a Purpose for Reading Think about these questions as you read:
- How does physical geography affect farming and fishing in North Africa, Southwest Asia, and Central Asia?
- What kinds of industries are important in the region?
- How are improvements in transportation and communications changing life in the region?

Main Idea

As you read pages 463–468 in your textbook, complete this graphic organizer by using the major headings of the section.

I. Meeting Food Needs

    A. _____

    B. _____

II. Industrial Growth

    A. _____

    B. _____

III. Transportation and Communications

    A. _____

    B. _____

    C. _____

    D. _____

IV. Interdependence

Meeting Food Needs (pages 463–465)

Reviewing

As you read the lesson, look for cause-and-effect relationships and then fill in the cause-and-effect chart below.
Review your chart when you have finished reading and
revise as needed.

Cause	Effect

Terms To Know

Define or describe the following key term from this lesson.

> **arable**

_____

_____

Terms To Review

Use the following terms that you studied earlier in a
sentence that reflects the term's meaning from this lesson.

> **declined**
> (Chapter 4, Section 1)

_____

_____

> **developed
> countries**
> (Chapter 4, Section 4)

_____

_____

Industrial Growth (pages 465–466)

Evaluating

As you read, form an opinion regarding the information in
this lesson. Does the quote by Erla Zwingle support the
following statement? Tourism is vital to Morocco. *Explain
your opinion.*

_____

_____

_____

Terms To Know

Write the letter of the correct definition in Group B next to the correct term from this lesson in Group A. One definition will not be used.

Group A

_____ **1.** commodities

_____ **2.** petrochemicals

_____ **3.** gross domestic product (GDP)

_____ **4.** hajj

Group B

a. economies based on oil

b. products that are made from petroleum or natural gas

c. pilgrimage

d. the value of goods and services produced in a country in a year

e. economic goods

Terms To Review

Use the following terms that you studied earlier in a sentence that reflects the term's meaning.

phosphate
(Chapter 17, Section 1)

_____

_____

_____

significant
(Chapter 18, Section 1)

_____

_____

Transportation and Communications (pages 466–468)

Analyzing

As you read this lesson, think about its organization and main ideas. Then write a sentence explaining the organization and list the main ideas.

Organization

_____

_____

_____

Main Ideas

_____

_____

_____

_____

Terms To Review

Use the following term that you studied earlier in a sentence that reflects the term's meaning from this lesson.

transport
(Chapter 8, Section 1)

_____

_____

Interdependence (page 468)

Inferring

As you read the lesson, look for clues in the descriptions and events that might help you draw a conclusion to the following question: **Why do oil prices rise and fall? How do these changes affect global consumers?**

_____

_____

Terms To Know

Define or describe the following key term from this lesson.

embargo

_____

_____

Academic Vocabulary

Define the following academic vocabulary words from this lesson.

aid

_____

_____

ensure

_____

_____

Terms To Review

Use the following term that you studied earlier in a sentence that reflects the term's meaning from this lesson.

despite
(Chapter 8, Section 1)

_____

_____

Section Wrap-up

Now that you have read the section, write the answers to the questions that were included in Setting a Purpose for Reading *at the beginning of the lesson.*

How does physical geography affect farming and fishing in North Africa, Southwest Asia, and Central Asia?

_____

_____

_____

What kinds of industries are important in the region?

_____

_____

_____

How are improvements in transportation and communications changing life in the region?

_____

_____

_____

Chapter 19, Section 2
People and Their Environment

(Pages 469–473)

Reason To Read

Setting a Purpose for Reading Think about these questions as you read:
• How have peoples in the region dealt with scarce water resources?
• What are the causes and effects of environmental problems in the region?

Main Idea

As you read pages 469–473 in your textbook, complete this graphic organizer by describing the environmental challenges of the Caspian Sea, Dead Sea, and Aral Sea.

Body of Water	Challenges
Caspian Sea	
Dead Sea	
Aral Sea	

The Need for Water *(pages 469–472)*

Interpreting

Think about the shortage of water in the region. Then ask yourself: What problems might occur if new sources of water are not found for the region? *Explain your answer.*

_____

_____

_____

_____

Terms To Know

Define or describe the following key terms from this lesson.

aquifers

_____

_____

desalination

_____

_____

Places To Locate

Explain why the following place from this lesson is important.

Tripoli

_____

_____

Terms To Review

Use the following terms that you studied earlier in a sentence that reflects the term's meaning from this lesson.

oases
(Chapter 3, Section 3)

_____

_____

layers
(Chapter 3, Section 3)

_____

_____

Environmental Concerns (pages 472–473)

Drawing Conclusions

As you read the lesson, think about the following questions: What are the effects of the Aswan High Dam? Have the effects of the dam been mostly positive or mostly negative?

Conclusion

_____
_____
_____
_____

Places To Locate

Explain why the following places from this lesson are important.

Aswan High Dam

_____
_____
_____

Elburz Mountains

_____
_____
_____

Dead Sea

_____
_____
_____

Aral Sea

_____
_____
_____

Terms To Review

Use the following terms that you studied earlier in a sentence that reflects the term's meaning from this lesson.

retains
(Chapter 3, Section 2)

_____
_____

project
(Chapter 10, Section 1)

_____
_____

 *Now that you have read the section, write the answers to the questions that were included in* Setting a Purpose for Reading *at the beginning of the lesson.*

How have peoples in the region dealt with scarce water resources?

_____

_____

_____

_____

What are the causes and effects of environmental problems in the region?

_____

_____

_____

_____

Chapter 20, Section 1
The Land

(Pages 499–504)

Reason To Read

Setting a Purpose for Reading Think about these questions as you read:
- What are the major landforms in Africa south of the Sahara?
- How does the land affect the water systems of Africa south of the Sahara?
- What are the region's most important natural resources?

Main Idea

As you read pages 499–504 in your textbook, complete this graphic organizer by filling in the many uses of Lake Volta.

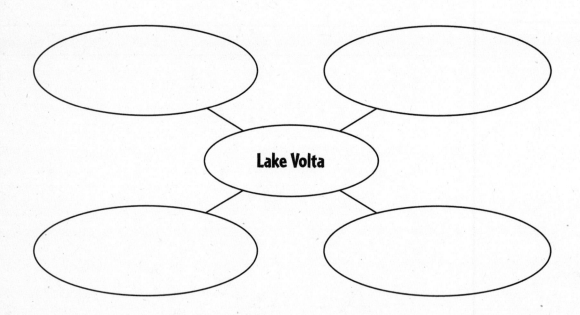

Landforms *(pages 499–500)*

Summarizing   *As you read, complete the following sentences. Doing so will help you summarize the lesson.*

1. Africa south of the Sahara is bounded in the north by the _____ , on the northeast by the _____ , on the west by the _____ , and to the east by the _____ .

2. The waters of the Atlantic and Indian Oceans meet at the _____ , on the southern tip of Africa.

3. Africa south of the Sahara is a series of _____ that rise in elevation from the coast inland.

Terms To Know   *Define or describe the following key terms from this lesson.*

escarpment ⟩ _____
_____

cataract ⟩ _____
_____

rift valley ⟩ _____
_____

fault ⟩ _____
_____

Chapter 20, Section 1   **241**

Places To Locate

Write the letter of the correct location in Group B next to the correct place from this lesson in Group A. One location will not be used.

Group A

_____ **1.** Ruwenzori Mountains

_____ **2.** Drakensberg Range

_____ **3.** Great Rift Valley

Group B

a. stretches from Syria in Southwest Asia to Mozambique

b. divides Uganda and the Democratic Republic of Congo

c. stretches from Ethiopia almost to the Cape of Good Hope

d. forms part of the escarpment along the southern edge of Africa

Academic Vocabulary

Define the following academic vocabulary word from this lesson.

parallel >

_____

_____

Terms To Review

Use the following terms that you studied earlier in a sentence that reflects the term's meaning from this lesson.

fjord
(Chapter 11, Section 1) >

_____

_____

_____

series
(Chapter 12, Section 2) >

_____

_____

Water Systems *(pages 500–504)*

Inferring

As you read the lesson, look for clues in the descriptions and events that might help you draw a conclusion to the question: **Why were Africans able to keep Europeans from gaining control of goods and trade routes in the interior of the continent?**

_____

_____

Terms To Know

Define or describe the following key terms from this lesson.

delta

_____
_____
_____

estuary

_____
_____

Places To Locate

Fill in the blank with the correct place from this lesson from the list below. One place will be used twice.

Lake Victoria

1. The _____ of south-central Africa originates near the Zambia-Angola border in the west and the Indian Ocean in the east.

Niger River

2. _____ lies between the eastern and western branches of the Great Rift Valley.

Zambezi River

3. The_____ originates in the highlands of Guinea and curves southeast to meet the Atlantic Ocean at the coast of Nigeria.

4. The source of the Nile River is _____ .

Academic Vocabulary

Define the following academic vocabulary word from this lesson.

capable

_____
_____

Terms To Review

Use the following term that you studied earlier in a sentence that reflects the term's meaning from this lesson.

generates
(Chapter 14, Section 1)

_____
_____

Natural Resources (page 504)

Previewing

Preview the lesson to get an idea of what is ahead. First, skim the lesson. Then write a sentence or two explaining what you think you will be learning. After you have finished reading, revise your statements as necessary.

_____

_____

_____

Terms To Review

Use the following term that you studied earlier in a sentence that reflects the term's meaning from this lesson.

hydroelectric power
(Chapter 8, Section 1)

_____

_____

Section Wrap-up

Now that you have read the section, write the answers to the questions that were included in Setting a Purpose for Reading *at the beginning of the lesson.*

What are the major landforms in Africa south of the Sahara?

_____

_____

How does the land affect the water systems of Africa south of the Sahara?

_____

_____

What are the region's most important natural resources?

_____

_____

Chapter 20, Section 2
Climate and Vegetation

(Pages 505–509)

Reason To Read

Setting a Purpose for Reading Think about these questions as you read:
• What geographic factors affect climate in Africa?
• What kinds of climate and vegetation are found in Africa south of the Sahara?

Main Idea

As you read pages 505–509 in your textbook, complete this graphic organizer by describing each geographical area.

Area	Description
Serengeti Plain	
Sahel	
Namib Desert	
Kalahari Desert	

Key Points

Notes

Tropical Climate *(pages 505–508)*

Visualizing

Visualize the information described in this lesson to help you understand and remember what you have read. First, read the lesson. Next, ask yourself, What would this look like? *Finally, write a description of the pictures you visualized on the lines below.*

_____

_____

_____

Terms To Know

Define or describe the following key terms from this lesson.

leach

_____

_____

savanna

_____

_____

harmattan

_____

_____

Places To Locate

Explain why the following place from this lesson is important.

Serengeti Plain

_____

_____

Academic Vocabulary

Define the following academic vocabulary word from this lesson.

diminish

_____

_____

Terms To Review

Use the following terms that you studied earlier in a sentence that reflects the term's meaning from this lesson.

clear-cutting
(Chapter 7, Section 2)

_____
_____
_____

deforestation
(Chapter 10, Section 2)

_____
_____

Dry Climates *(pages 508–509)*

Previewing

Read the title and main headings of the lesson. Write a statement predicting what the lesson is about and what will be included in the text. As you read, adjust or change your prediction if it does not match what you learn.

_____
_____
_____

Places To Locate

Explain why the following places from this lesson are important.

Sahel

_____
_____

Namib Desert

_____
_____

Kalahari Desert

_____
_____

Terms To Review

Use the following term that you studied earlier in a sentence that reflects the term's meaning from this lesson.

eroded
(Chapter 5, Section 2)

_____

_____

Moderate Climates *(page 509)*

Evaluating

As you read, form an opinion about the information in this lesson. Answer the following questions:

1. Is Curt Stager a reliable source? Explain your answer.

_____

_____

2. Does the quote by Curt Stager support the following statement? *The highlands areas can seem almost lush.* Explain your opinion.

_____

_____

Section Wrap-up

Now that you have read the section, write the answers to the questions that were included in Setting a Purpose for Reading *at the beginning of the lesson.*

What geographic factors affect climate in Africa?

_____

_____

What kinds of climate and vegetation are found in Africa south of the Sahara?

_____

_____

Chapter 21, Section 1
Population Patterns

(Pages 515–518)

Reason To Read

Setting a Purpose for Reading Think about these questions as you read:
- Why are parts of Africa south of the Sahara densely populated?
- What are the obstacles to economic growth in the region?
- Who are the diverse peoples of Africa south of the Sahara?
- Why are the region's cities growing so rapidly?

Main Idea

As you read pages 515–518 in your textbook, complete this graphic organizer by filling in reasons for food production problems in Africa south of the Sahara.

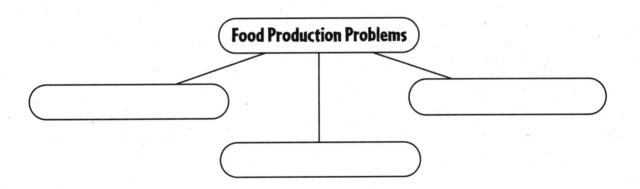

Rapid Population Growth (pages 515–517)

Synthesizing

As you read, think about the main ideas of the lesson. Use the main ideas to answer the following question.

How can Africa south of the Sahara have rapid population growth and yet have relatively few people in relation to its vast land area?

_____

_____

_____

Terms To Know

Define or describe the following key term from this lesson.

sanitation

_____

_____

Places To Locate

Explain why the following places from this lesson are important.

Nigeria

_____

_____

_____

Rwanda

_____

_____

_____

Namibia

_____

_____

_____

Zimbabwe

_____

_____

_____

Academic Vocabulary

Define the following academic vocabulary words from this lesson.

disposal

_____

_____

Key Points

Notes

transmit >

_____
_____

Terms To Review

Use the following terms that you studied earlier in a sentence that reflects the term's meaning from this lesson.

death rate
(Chapter 4, Section 1) >

_____
_____
_____

estimated
(Chapter 9, Section 1) >

_____
_____

A Diverse Population *(pages 517–518)*

Previewing

Preview the lesson to get an idea of what is ahead. First, skim the lesson. Then write a sentence or two explaining what you think you will be learning. After you have finished reading, revise your statements as necessary.

_____
_____
_____

Terms To Know

Define or describe the following key terms from this lesson.

urbanization >

_____
_____

service centers >

_____
_____

Terms To Review

Use the following terms that you studied earlier in a sentence that reflects the term's meaning from this lesson.

ethnic groups
(Chapter 4, Section 2)

_____

_____

diverse
(Chapter 6, Section 1)

_____

_____

Section Wrap-up

Now that you have read the section, write the answers to the questions that were included in Setting a Purpose for Reading *at the beginning of the lesson.*

Why are parts of Africa south of the Sahara densely populated?

_____

_____

What are the obstacles to economic growth in the region?

_____

_____

Who are the diverse peoples of Africa south of the Sahara?

_____

_____

Why are the region's cities growing so rapidly?

_____

_____

Chapter 21, Section 2
History and Government

(Pages 519–524)

Reason To Read

Setting a Purpose for Reading Think about these questions as you read:
- What were the main achievements of the ancient civilizations of Africa south of the Sahara?
- How did European colonization disrupt African patterns of life?
- What challenges did countries of the region face after independence?

Main Idea

As you read pages 519–524 in your textbook, complete this graphic organizer by describing the ancient trading empires.

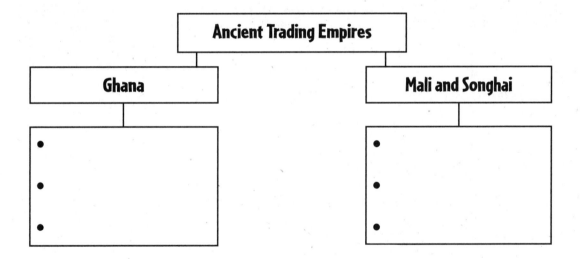

African Roots (pages 519–521)

Drawing Conclusions

As you read the lesson, look for clues to help you answer the following question: Why do you think the West African trading kingdoms were willing to trade gold for salt?

Conclusion ⟩ _____

_____

Terms To Know

Define or describe the following key term from this lesson.

domesticate ⟩ _____

_____

Places To Locate

Fill in the blank with the correct place from this lesson from the list below. One place will be used twice.

Kush ⟩

1. _____ was the center of the Mali empire.

2. _____ was a trading empire that developed in West Africa about A.D. 700.

Axum ⟩

3. The kingdom of _____ was located in what is now Sudan, and extended north into Egyptian territory.

Ghana ⟩

4. The _____ empire took over the Mali empire in West Africa and extended east.

Mali ⟩

5. _____ was a powerful trading empire in northern Ethiopia.

Songhai ⟩

6. The West African trading empire of _____ extended west to the Atlantic Ocean and was larger than Egypt.

Timbuktu ⟩

7. The _____ empire succeeded the Ghana empire in West Africa.

 Key Points

 Notes

Terms To Review

Use the following terms that you studied earlier in a sentence that reflects the term's meaning from this lesson.

debated
(Chapter 10, Section 2)

_____

_____

_____

shift
(Chapter 17, Section 1)

_____

_____

_____

European Colonization (pages 521–523)

Sequencing

As you read, number the following events in the correct order.

_____ Europeans begin trade with Africans.

_____ The king of Kongo complains about Portuguese slave merchants.

_____ Most of Africa is under European control.

_____ Arab traders begin bringing enslaved Africans to the Islamic world.

_____ Europeans begin to claim African territory.

Places To Locate

Explain why the following places from this lesson are important.

Ethiopia

_____

_____

_____

Liberia

_____

_____

_____

South Africa

_____

_____

_____

Key Points

Notes

Academic Vocabulary

Circle the letter of the word or phrase that has the closest meaning to the underlined word from this lesson.

The king of Kongo, in statements <u>quoted</u> in *African Kingdoms*, complained to the king of Portugal about Portuguese slave merchants.

a. repeated   **b.** challenged   **c.** mistreated

Terms To Review

Use the following terms that you studied earlier in a sentence that reflects the term's meaning from this lesson.

source
(Chapter 2, Section 3)

_____

_____

focused
(Chapter 12, Section 3)

_____

_____

From Colonies to Countries *(pages 523–524)*

Synthesizing

As you read, think about the main ideas of the lesson. Combine the ideas in this lesson to answer the following question.

In what ways did colonialism affect the region's development and set the stage for current conflicts in Africa south of the Sahara?

_____

_____

_____

_____

Terms To Know

Define or describe the following key terms from this lesson.

apartheid

_____

_____

universal suffrage

_____

Academic Vocabulary

Circle the letter of the word or phrase that has the closest meaning to the underlined word from this lesson.

1. The South African government <u>imposed</u> apartheid on South Africa's black majority and racially mixed peoples.

   a. discouraged   **b.** forced   **c.** stopped

2. Under apartheid, nonwhite South Africans were <u>denied</u> political rights and equality with whites.

   a. guaranteed   **b.** told about   **c.** not given

Terms To Review

Use the following terms that you studied earlier in a sentence that reflects the term's meaning from this lesson.

democracy
(Chapter 4, Section 3)

_____

_____

policy
(Chapter 13, Section 1)

_____

_____

Section Wrap-up

Now that you have read the section, write the answers to the questions that were included in Setting a Purpose for Reading *at the beginning of the lesson.*

What were the main achievements of the ancient civilizations of Africa south of the Sahara?

_____

_____

How did European colonization disrupt African patterns of life?

_____

_____

What challenges did countries of the region face after independence?

_____

_____

Chapter 21, Section 3
Cultures and Lifestyles
(Pages 525–529)

Reason To Read

Setting a Purpose for Reading Think about these questions as you read:
- What languages do people in Africa south of the Sahara speak?
- What are the major religions in Africa south of the Sahara?
- What art forms have peoples of the region developed?
- How do lifestyles among peoples of the region differ? How are they similar?

Main Idea

As you read pages 525–529 in your textbook, complete this graphic organizer by filling in the common elements of traditional religions in the region.

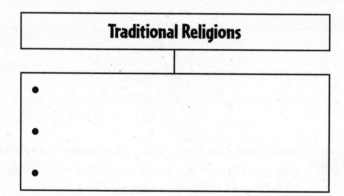

Traditional Religions

-
-
-

Languages *(pages 525–526)*

Scanning

Scan the lesson before you begin to read. As you glance quickly over the lines of text, look for key words or phrases that will tell you what the text will cover. Write the key words or phrases. Then use the key words and phrases to write a statement explaining the lesson content. Revise your statement when you are finished reading the lesson.

Key words or phrases >
_____
_____
_____

What the lesson is about >
_____
_____

Terms To Know

Define or describe the following key terms from this lesson.

mass culture >
_____
_____
_____

lingua franca >
_____
_____

Academic Vocabulary

Circle the letter of the word or phrase that has the closest meaning to the underlined word from this lesson.

1. Language experts put the many ethnic groups and languages of Africa south of the Sahara into six major <u>categories</u>.

    a. containers   **b.** stages   **c.** divisions

2. Afrikaans contains words <u>adapted</u> from English, French, German, and African languages.

    a. omitted   **b.** borrowed   **c.** rejected

Terms To Review

Use the following terms that you studied earlier in a sentence that reflects the term's meaning from this lesson.

dialect
(Chapter 9, Section 1)

_____

_____

_____

experts
(Chapter 12, Section 1)

_____

_____

Religions (pages 526–527)

Determining the Main Idea

As you read, write the main idea of the lesson. Review your statement when you have finished reading and revise as needed.

_____

_____

Academic Vocabulary

Circle the letter of the word or phrase that has the closest meaning to the underlined word from this lesson.

1. Religion plays an <u>integral</u> role in everyday life in Africa.

    a. interesting   **b.** unimportant   **c.** essential, complete

Education (page 527)

Connecting

As you read, compare the educational systems of Africa south of the Sahara to that of the United States. Summarize your thoughts in a paragraph. Be sure to include ways that education in this region is similar to and different from education in the United States.

_____

_____

_____

_____

Terms To Review

Use the following terms that you studied earlier in a sentence that reflects the term's meaning from this lesson.

available
(Chapter 2, Section 3)

_____
_____
_____

literacy rates
(Chapter 6, Section 3)

_____
_____
_____

The Arts *(pages 527–529)*

Inferring

As you read the lesson, look for clues in the descriptions and events that might help you draw a conclusion. Use these clues to answer the following question.

Why do you think storytellers are respected figures in African communities?

_____
_____

Terms To Know

Define or describe the following key term from this lesson.

oral tradition

_____
_____

Academic Vocabulary

Define the following academic vocabulary words from this lesson.

medium

_____
_____

specific

_____
_____

Terms To Review

Use the following term that you studied earlier in a sentence that reflects the term's meaning from this lesson.

distinct
(Chapter 11, Section 1)

_____

_____

Varied Lifestyles *(page 529)*

Connecting

As you read, compare the lifestyles of people in Africa south of the Sahara with the lifestyles of people in the United States. Summarize your thoughts in a paragraph. Be sure to include ways that the region's lifestyles are similar to and different from American lifestyles.

_____

_____

_____

Terms To Know

Write the letter of the correct definition in Group B next to the correct term from this lesson in Group A. One definition will not be used.

Group A

_____ **1.** extended families

_____ **2.** clans

_____ **3.** nuclear families

Group B

a. households made up of husband, wife, and children

b. households made up of several generations

c. large groups of people descended from an early common ancestor

d. ethnic groups who live in a country

Academic Vocabulary

Define the following academic vocabulary word from this lesson.

generations

_____

_____

Section Wrap-up

Now that you have read the section, write the answers to the questions that were included in Setting a Purpose for Reading *at the beginning of the lesson.*

What languages do people in Africa south of the Sahara speak?

_____

_____

_____

What are the major religions in Africa south of the Sahara?

_____

_____

_____

What art forms have peoples of the region developed?

_____

_____

_____

How do lifestyles among peoples of the region differ? How are they similar?

_____

_____

_____

Chapter 22, Section 1
Living in Africa South of the Sahara

(Pages 537–542)

Reason To Read

Setting a Purpose for Reading Think about these questions as you read:

- What are the most common farming methods in Africa south of the Sahara?
- How do mineral resources benefit the peoples of the region?
- Why has industrial development been slow in Africa south of the Sahara?
- How are transportation and communications in the region changing?

Main Idea

As you read pages 537–542 in your textbook, complete this graphic organizer by filling in the obstacles farmers face in Africa south of the Sahara.

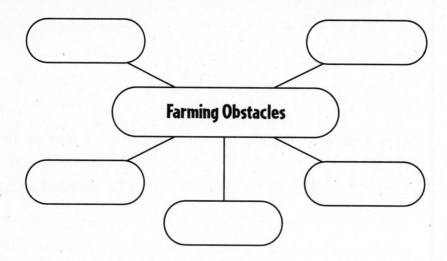

Farming Obstacles

Agriculture (pages 537–539)

Skimming

Read the title and quickly look over the lesson to get a general idea of the lesson's content. Then write a sentence or two explaining what the lesson is about.

_____

_____

_____

_____

Terms To Know

Write the letter of the correct definition in Group B next to the correct term from this lesson in Group A. One definition will not be used.

Group A

_____ **1.** subsistence farming

_____ **2.** shifting farming

_____ **3.** sedentary farming

_____ **4.** commercial farming

_____ **5.** cash crops

_____ **6.** conservation farming

Group B

a. farm crops grown to be sold or traded rather than used by the farm family

b. producing just enough food for a family or village to survive

c. agriculture conducted at permanent settlements

d. single-crop economies

e. a method in which farmers move every one to three years to find better soil

f. a land-management technique that helps protect farmland

g. farms produce crops on a large scale

Academic Vocabulary

Define the following academic vocabulary words from this lesson.

instance _____

_____

techniques _____

_____

Logging and Fishing (pages 539–540)

Monitoring Comprehension

As you read the lesson, write down questions you have about what you read. When you have finished reading the lesson, answer your questions. If you cannot answer your questions, ask your teacher for clarification.

_____

_____

_____

_____

Mining Resources (page 540)

Responding

As you read the lesson, think about what attracts your attention as you read. Write down facts you find interesting or surprising in the lesson.

_____

_____

_____

_____

Academic Vocabulary

Circle the letter of the word or phrase that has the closest meaning to the underlined word from this lesson.

South African miners <u>extract</u> large quantities of coal, platinum, chromium, vanadium, and manganese for export.

a. remove   **b.** find   **c.** sell

Industrialization (pages 540–541)

Analyzing

As you read this lesson, think about its organization and main ideas. Then write a sentence explaining the organization and list the main ideas.

Organization >

_____

_____

_____

Main Idea >

_____

_____

_____

Terms To Know

Define or describe the following key term from this lesson.

infrastructure >

_____

_____

Academic Vocabulary

Define the following academic vocabulary word from this lesson.

assemble >

_____

_____

Transportation and Communications (pages 541–542)

Interpreting

Think about the uses of the Internet. Ask yourself, how does the Internet broaden the market for locally made products in the region?

_____

_____

 **Key Points**

 **Notes**

Terms To Know

Define or describe the following key term from this lesson.

e-commerce

_____

_____

Academic Vocabulary

Define the following academic vocabulary word from this lesson.

scheduling

_____

_____

Section Wrap-up

Now that you have read the section, write the answers to the questions that were included in Setting a Purpose for Reading *at the beginning of the lesson.*

What are the most common farming methods in Africa south of the Sahara?

_____

_____

How do mineral resources benefit the peoples of the region?

_____

_____

Why has industrial development been slow in Africa south of the Sahara?

_____

_____

How are transportation and communications in the region changing?

_____

_____

Chapter 22, Section 2
People and Their Environment

(Pages 543–547)

Reason To Read

Setting a Purpose for Reading Think about these questions as you read:
- Why have food shortages occurred in parts of Africa south of the Sahara?
- What steps are the African countries south of the Sahara taking to protect their environment?
- What is the outlook for the region's future development?

Main Idea

As you read pages 543–547 in your textbook, complete this graphic organizer by describing the farming methods used in the region.

Farming Methods	Description
Subsistence	
Shifting	
Sedentary	
Commercial	
Conservation	

Shadow of Hunger *(pages 543–545)*

Reviewing

As you read, list the actions that governments in Africa south of the Sahara have taken to help eliminate hunger. Then review the listed actions and rank them by order of importance. Explain your reasoning.

_____

_____

_____

_____

Terms To Review

Use the following term that you studied earlier in a sentence that reflects the term's meaning from this lesson.

refugees
(Chapter 12, Section 1)

_____

_____

Land Use *(pages 545–547)*

Clarifying

As you read this lesson, write down terms or concepts you find confusing. Then go back and reread the lesson to clarify the confusing parts. Write an explanation of the terms and concepts. If you are still confused, ask your teacher to clarify.

_____

_____

_____

_____

Terms To Know

Fill in the blank with the correct term from this lesson from the list below. One term will be used twice.

habitats   extinction   poaching   ecotourism

1. A big business in the region is _____ , or tourism based on concern for the environment.

2. Hundreds of animals in the region that exist now are in danger of

_____ , or disappearance from the earth.

3. The living areas, or _____ , of many plants and
animals are being destroyed by deforestation.

4. Huge game reserves in the region have helped keep many endangered

animals from _____ .

5. Fewer than 600,000 elephants remain in the region because of

_____ , or illegal hunting.

Academic Vocabulary

Define the following academic vocabulary words from this lesson.

statistics >

_____

_____

_____

pursue >

_____

_____

Terms To Review

Use the following term that you studied earlier in a sentence that reflects the term's meaning from this lesson.

impact >
(Chapter 7, Section 2)

_____

_____

Toward the Future *(page 547)*

Drawing Conclusions

As you read the lesson, think about the following question:
**How have African countries addressed the problem of
endangered species?**

_____

_____

Academic Vocabulary

Define the following academic vocabulary word from this lesson.

positive

_____

_____

Section Wrap-up

Now that you have read the section, write the answers to the questions that were included in Setting a Purpose for Reading *at the beginning of the lesson.*

Why have food shortages occurred in parts of Africa south of the Sahara?

_____

_____

What steps are the African countries south of the Sahara taking to protect their environment?

_____

_____

What is the outlook for the region's future development?

_____

_____

Chapter 23, Section 1
The Land

(Pages 569–574)

Reason To Read

Setting a Purpose for Reading Think about these questions as you read:
- What landforms exist in South Asia?
- What are the three great river systems on which life in South Asia depends?
- How do the peoples of South Asia use the region's natural resources?

Main Idea

As you read pages 569–574 in your textbook, complete this graphic organizer by describing the region's three major river systems.

River System	Description
Indus River	
Brahmaputra	
Ganges River	

A Separate Land (page 569)

Summarizing

As you read, complete the following sentences. Doing so will help you summarize the lesson.

1. South Asia is separated from the rest of Asia by _____.

2. Most of South Asia forms a _____ of about 1.7 million square miles (4.4 million sq. km).

3. South Asia borders three bodies of water—the _____ to the west, the _____ to the south, and the _____ to the east.

4. South Asia's large island country of _____ lies off India's southern tip.

Terms To Know

Define or describe the following key term from this lesson.

subcontinent ⟩ _____

_____

A Land of Great Variety (pages 570–571)

Synthesizing

As you read the lesson, look for clues in the descriptions and events that might help you answer the following questions: **How does the landscape of the Himalaya differ from that of the Deccan Plateau? How do these differences affect people's lives?**

_____

_____

_____

_____

Chapter 23, Section 1

Academic Vocabulary

Define the following academic vocabulary word from this lesson.

chapter >
_____
_____

Terms To Review

Use the following terms that you studied earlier in a sentence that reflects the term's meaning from this lesson.

continental drift
(Chapter 2, Section 2) >
_____
_____

theory
(Chapter 6, Section 2) >
_____
_____

Major River Systems (page 572)

Visualizing

Visualize the information described is this lesson to help you understand and remember what you have read. First, read the lesson. Next, ask yourself, **What would one of the major river systems of South Asia look like?** *Finally, write a description of the pictures you visualized on the lines below.*

_____
_____
_____

Terms To Know

Define or describe the following key term from this lesson.

alluvial plain >
_____
_____

 **Key Points**

 **Notes**

Terms To Review

Use each of the following terms that you studied earlier in a sentence that reflects the term's meaning from this lesson.

retains
(Chapter 3, Section 2)

_____
_____
_____

delta
(Chapter 20, Section 1)

_____
_____

Natural Resources *(pages 573–574)*

Previewing

Preview the lesson to get an idea of what is ahead. First, skim the lesson. Then write a sentence or two explaining what you think you will be learning. After you have finished reading, revise your statements as necessary.

_____
_____
_____

Terms To Know

Define or describe the following key term from this lesson.

mica

_____
_____

Academic Vocabulary

Define the following academic vocabulary word from this lesson.

ongoing

_____
_____

 **Key Points**

 **Notes**

**Terms To Review**

Use the following terms that you studied earlier in a sentence that reflects the term's meaning from this lesson.

overall
(Chapter 4, Section 1)

_____

_____

funded
(Chapter 15, Section 2)

_____

_____

Section Wrap-up

Now that you have read the section, write the answers to the questions that were included in Setting a Purpose for Reading *at the beginning of the lesson.*

What landforms exist in South Asia?

_____

_____

What are the three great river systems on which life in South Asia depends?

_____

_____

How do the peoples of South Asia use the region's natural resources?

_____

_____

Chapter 23, Section 2
Climate and Vegetation

(Pages 575–579)

Reason To Read

Setting a Purpose for Reading Think about these questions as you read:
- What are the five major climate regions of South Asia?
- How do seasonal weather patterns present challenges to the region's economy?
- How do elevation and rainfall affect South Asia's vegetation?

Main Idea

As you read pages 575–579 in your textbook, complete this graphic organizer by identifying the three seasons that occur in the region.

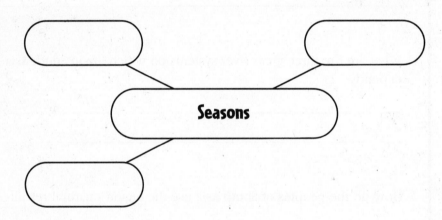

Seasons

South Asia's Climates (pages 575–577)

Scanning

Scan the lesson before you begin to read. As you glance quickly over the lines of text, look for key words or phrases that will tell you what the text will cover. Write the key words or phrases. Then use the key words and phrases to write a statement explaining the lesson content. Revise your statement when you are finished reading the lesson.

Key words or phrases

_____
_____
_____
_____

What the lesson is about

_____
_____
_____
_____

Places To Locate

Explain why the following places from this lesson are important.

Bay of Bengal

_____
_____
_____

Great Indian Desert

_____
_____
_____

Terms To Review

Use the following terms that you studied earlier in a sentence that reflects the term's meaning from this lesson.

annual
(Chapter 3, Section 2)

_____
_____
_____

steppe
(Chapter 14, Section 2)

_____
_____
_____

Monsoons (pages 577–579)

Predicting

Read the title and main headings of the lesson. Write a statement predicting what the lesson is about and what will be included in the text. As you read, adjust or change your prediction if it does not match what you learn.

_____

_____

_____

Terms To Know

Define or describe the following key terms from this lesson.

monsoons

_____

_____

cyclone

_____

_____

Academic Vocabulary

Define the following academic vocabulary word from this lesson.

approaching

_____

_____

Terms To Review

Use the following terms that you studied earlier in a sentence that reflects the term's meaning from this lesson.

evaporation
(Chapter 2, Section 3)

_____

_____

shifted
(Chapter 17, Section 1)

_____

_____

Now that you have read the section, write the answers to the questions that were included in Setting a Purpose for Reading *at the beginning of the lesson.*

What are the five major climate regions of South Asia?

_____

_____

How do seasonal weather patterns present challenges to the region's economy?

_____

_____

How do elevation and rainfall affect South Asia's vegetation?

_____

_____

Chapter 24, Section 1
Population Patterns
(Pages 587–591)

Reason To Read

Setting a Purpose for Reading Think about these questions as you read:
- How do peoples of South Asia reflect diversity?
- How is South Asia's large population distributed?
- How does life in the region's cities compare with life in traditional rural villages?

Main Idea

As you read pages 587–591 in your textbook, complete this graphic organizer by describing India's main cities.

City	Description
Mumbai (Bombay)	
Kolkata (Calcutta)	
Delhi	

Human Characteristics (pages 587–588)

Synthesizing

As you read, think about the main ideas of the lesson. Combine the ideas in this lesson to answer the following question.

Would you say that diversity has been more of a problem or a benefit for countries in South Asia? Why?

_____

_____

_____

_____

_____

Terms To Know

Define or describe the following key term from this lesson.

jati

_____

_____

Places To Locate

Explain why the following place from this lesson is important.

Islamabad

_____

_____

Terms To Review

Use the following terms that you studied earlier in a sentence that reflects the term's meaning from this lesson.

identify
(Chapter 2, Section 3)

_____

_____

complex
(Chapter 16, Section 1)

_____

_____

Population Density and Distribution (pages 588–590)

Responding

As you read this lesson, think about what attracts your attention as you read. Write down facts you find interesting or surprising in the lesson.

_____

_____

_____

_____

Terms To Review

Use the following terms that you studied earlier in a sentence that reflects the term's meaning from this lesson.

population density
(Chapter 4, Section 1)

_____

_____

exceed
(Chapter 10, Section 2)

_____

_____

Urban and Rural Life (pages 590–591)

Previewing

Preview the lesson to get an idea of what is ahead. First, skim the lesson. Then write a sentence or two explaining what you think you will be learning. After you have finished reading, revise your statements as necessary.

_____

_____

_____

Terms To Know

Define or describe the following key term from this lesson.

megalopolis

_____

_____

Terms To Review

Use the following terms that you studied earlier in a sentence that reflects the term's meaning from this lesson.

clans
(Chapter 21, Section 3)

_____

_____

_____

extended families
(Chapter 9, Section 3)

_____

_____

Section Wrap-up

Now that you have read the section, write the answers to the questions that were included in Setting a Purpose for Reading *at the beginning of the lesson.*

How do peoples of South Asia reflect diversity?

_____

_____

How is South Asia's large population distributed?

_____

_____

How does life in the region's cities compare with life in traditional rural villages?

_____

_____

Chapter 24, Section 2
History and Government
(Pages 592–597)

Reason To Read

Setting a Purpose for Reading Think about these questions as you read:
- Where did South Asia's first civilization develop?
- What two major world religions originated in South Asia?
- How did invasions and conquests shape South Asia?
- What type of challenges are South Asian countries facing today?

Main Idea

As you read pages 592–597 in your textbook, complete this graphic organizer by using the headings of the section to create an outline.

I. Early History

    A. _____

    B. _____

II. Two Great Religions

    A. _____

    B. _____

    C. _____

III. Invasions and Empires

    A. Mauryan Empire _____

    B. Gupta Empire _____

IV. Modern South Asia

    A. _____

    B. _____

    C. _____

Early History (pages 592–593)

Drawing Conclusions

As you read the lesson, look for clues to help you answer the following question: What environmental changes led to the decline of the Indus Valley civilization?

Conclusion _____

_____

Academic Vocabulary

Circle the letter of the word or phrase that has the closest meaning to the underlined word from this lesson.

At first, the boundaries between social groups in the Indus Valley civilization were <u>flexible</u>, but gradually the social structure became more rigid.

a. lenient   **b.** important   **c.** strict

Terms To Review

Use the following term that you studied earlier in a sentence that reflects the term's meaning from this lesson.

sanitation
(Chapter 21, Section 1)

_____

_____

Two Great Religions (pages 593–594)

Synthesizing

As you read, think about the main ideas of the lesson. Combine the ideas in this lesson to answer the following question.

How does Hinduism regard other religions?

_____

_____

_____

_____

Terms To Know

Write the letter of the correct definition in Group B next to the correct term from this lesson in Group A. One definition will not be used.

Group A

____ **1.** dharma

____ **2.** reincarnation

____ **3.** karma

____ **4.** nirvana

Group B

a. the true nature of human existence

b. ultimate state of peace, calm, joy, and insight toward which people strive

c. a person's moral duty, based on class distinctions, which guides his or her life

d. the sum of good and bad actions in one's present and past lives

e. being born repeatedly in different forms, until one has overcome earthly desires

Academic Vocabulary

Define the following academic vocabulary words from this lesson.

insights

_____

_____

_____

rejected

_____

_____

Terms To Review

Use the following terms that you studied earlier in a sentence that reflects the term's meaning from this lesson.

cycle
(Chapter 3, Section 1)

_____

_____

temporary
(Chapter 9, Section 1)

_____

_____

Invasions and Empires (pages 594–595)

Sequencing

As you read, number the following events in the correct order.

_____ Gupta Empire comes into power in South Asia.

_____ Mauryan Empire gains control of South Asia.

_____ Muslim missionaries and traders first enter India.

_____ Europeans first arrive in South Asia.

_____ Muslim armies conquer northern India.

Terms To Know

Define or describe the following key term from this lesson.

raj

_____

_____

Terms To Review

Use the following term that you studied earlier in a sentence that reflects the term's meaning from this lesson.

elements
(Chapter 2, Section 2)

_____

_____

Modern South Asia (pages 595–597)

Evaluating

When you evaluate something you read, you make a judgment or form an opinion. After you read this lesson, decide whether or not you agree with Gandhi's philosophy of using nonviolent methods to seek self-rule and explain why.

_____

_____

_____

_____

Key Points

Notes

Academic Vocabulary

Define the following academic vocabulary words from this lesson.

> rigid

_____
_____
_____

> stable

_____
_____

Terms To Review

Use the following term that you studied earlier in a sentence that reflects the term's meaning from this lesson.

> methods
> (Chapter 1, Section 2)

_____
_____

Section Wrap-up

Now that you have read the section, write the answers to the questions that were included in **Setting a Purpose for Reading** *at the beginning of the lesson.*

Where did South Asia's first civilization develop?

_____
_____

What two major world religions originated in South Asia?

_____
_____

How did invasions and conquests shape South Asia?

_____
_____
_____

What type of challenges are South Asian countries facing today?

_____
_____

Chapter 24, Section 3
Cultures and Lifestyles
(Pages 600–605)

Reason To Read

Setting a Purpose for Reading Think about these questions as you read:
- How do the lives of South Asia's people reflect the region's linguistic and religious diversity?
- What contributions to the arts has the region made?
- How are South Asian countries meeting challenges to improve the quality of life in the region?
- How is the rich cultural diversity of South Asia reflected in distinctive celebrations?

Main Idea

As you read pages 600–605 in your textbook, complete this graphic organizer by filling in the three Indo-European languages.

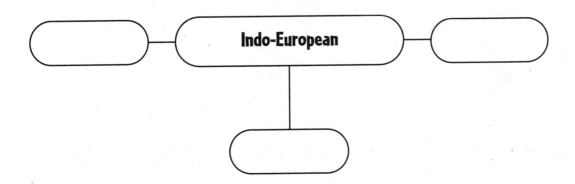

Key Points

Notes

Languages *(pages 600–601)*

Scanning

Scan the lesson before you begin to read. As you glance quickly over the lines of text, look for key words or phrases that will tell you what the text will cover. Write the key words or phrases. Then use the key words and phrases to write a statement explaining the lesson content. Revise your statement when you are finished reading the lesson.

Key words or phrases ⟩

_____

_____

_____

_____

What the lesson is about ⟩

_____

_____

_____

Terms To Review

Use the following term that you studied earlier in a sentence that reflects the term's meaning from this lesson.

trace
(Chapter 9, Section 3) ⟩

_____

_____

Religions *(pages 601–602)*

Determining the Main Idea

As you read, write the main idea of the lesson. Review your statement when you have finished reading and revise as needed.

_____

_____

Terms To Know

Define or describe the following key terms from this lesson.

guru ⟩

mantras	
	_____
	_____
	_____
sadhus	_____
	_____

The Arts *(pages 602–604)*

Predicting

Read the title and main headings of the lesson. Write a statement predicting what the lesson is about and what will be included in the text. As you read, adjust or change your prediction if it does not match what you learn.

_____

_____

_____

Terms To Know

Define or describe the following key terms from this lesson.

stupas	_____
	_____
dzong	_____
	_____

Academic Vocabulary

Define the following academic vocabulary words from this lesson.

drama	_____
	_____
	_____
contemporary	_____
	_____

Key Points

Notes

Terms To Review

Use the following terms that you studied earlier in a sentence that reflects the term's meaning from this lesson.

series
(Chapter 12, Section 2)

_____

_____

techniques
(Chapter 22, Section 1)

_____

_____

Quality of Life *(pages 604–605)*

Inferring

As you read the lesson, look for clues in the descriptions and events that might you draw a conclusion. Use these clues to answer the following question.

What improvements in health and education might enrich the quality of life in South Asia?

_____

_____

Academic Vocabulary

Define the following academic vocabulary word from this lesson.

nevertheless

_____

_____

Celebrations *(page 605)*

Connecting

As you read, compare the celebrations of South Asia with the celebrations of the United States. Summarize your thoughts in a paragraph.

_____

_____

_____

_____

Section Wrap-up

Now that you have read the section, write the answers to the questions that were included in **Setting a Purpose for Reading** *at the beginning of the lesson.*

How do the lives of South Asia's people reflect the region's linguistic and religious diversity?

_____

_____

What contributions to the arts has the region made?

_____

_____

How are South Asian countries meeting challenges to improve the quality of life in the region?

_____

_____

How is the rich cultural diversity of South Asia reflected in distinctive celebrations?

_____

_____

Chapter 25, Section 1
Living in South Asia

(Pages 611–617)

Reason To Read

Setting a Purpose for Reading Think about these questions as you read:

• How does agriculture provide a living for most of South Asia's people?
• What role do fisheries and mines have in South Asian economies?
• Where in South Asia is rapid industrial development taking place?
• What issues are raised by tourism in South Asia?

Main Idea

As you read pages 611–617 in your textbook, complete this graphic organizer by filling in the obstacles new farming methods create for the people of South Asia.

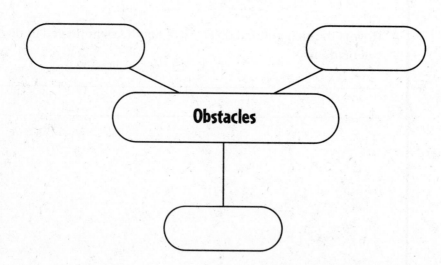

Living From the Land (pages 611–612)

Skimming

Read the title and quickly look over the lesson to get a general idea of the lesson's content. Then write a sentence or two explaining what the lesson will cover.

_____

_____

_____

Terms To Review

Use the following terms that you studied earlier in a sentence that reflects the term's meaning from this lesson.

delta
(Chapter 20, Section 1)

_____

_____

subsistence farming
(Chapter 22, Section 1)

_____

_____

South Asian Crops (pages 612–613)

Monitoring Comprehension

As you read the lesson, write down questions you have about what you read. When you have finished reading the lesson, answer your questions.

_____

_____

_____

_____

Terms To Know

Write the letter of the correct definition in Group B next to the correct term from this lesson in Group A. One definition will not be used.

Group A

_____ **1.** cash crops

_____ **2.** jute

_____ **3.** green revolution

_____ **4.** biomass

Group B

a. farm crops grown for sale or export

b. plant materials and animal dung

c. a fiber used to make string, rope, and cloth

d. agriculture conducted at permanent settlements

e. a movement to increase and diversify crop yields in developing countries

Terms To Review

Use the following terms that you studied earlier in a sentence that reflects the term's meaning.

research
(Chapter 1, Section 2)

_____

_____

cycles
(Chapter 3, Section 1)

_____

_____

Mining and Fishing (pages 613–614)

Responding

As you read this lesson, think about what attracts your attention as you read. Write down facts you find interesting or surprising in the lesson.

_____

_____

_____

_____

Academic Vocabulary

Circle the letter of the word or phrase that has the closest meaning to the underlined word from this lesson.

Bangladesh may use natural gas to <u>supplement</u> the country's export income because of the declining market for jute.

a. replace   **b.** add to   **c.** sell

Terms To Review

Use the following term that you studied earlier in a sentence that reflects the term's meaning from this lesson.

fisheries
(Chapter 5, Section 1)

_____

_____

South Asian Industries *(pages 614–616)*

Analyzing

As you read this lesson, think about its organization and main ideas. Then write a sentence explaining the organization and list three main ideas.

Organization

_____

_____

Main Ideas

_____

_____

_____

Terms To Know

Define or describe the following key term from this lesson.

cottage industries

_____

_____

Academic Vocabulary

Define the following academic vocabulary words from this lesson.

styles

_____

_____

professional

Tourism (pages 616–617)

Interpreting

Think about tourism in South Asia. Ask yourself, how might increased tourism affect life in the region?

_____

_____

Terms To Know

Define or describe the following key term from this lesson.

ecotourism

_____

_____

Section Wrap-up

Now that you have read the section, write the answers to the questions that were included in Setting a Purpose for Reading *at the beginning of the lesson.*

How does agriculture provide a living for most of South Asia's people?

_____

_____

What role do fisheries and mines have in South Asian economies?

_____

_____

Where in South Asia is rapid industrial development taking place?

_____

_____

What issues are raised by tourism in South Asia?

_____

_____

Chapter 25, Section 2
People and Their Environment

(Pages 618–623)

Reason To Read

Setting a Purpose for Reading Think about these questions as you read:
- How is South Asia handling the complex task of managing its rich natural resources?
- What environmental challenges does South Asia face in the years ahead?
- How do geographic factors impact the political and economic challenges of South Asia's future?

Main Idea

As you read pages 618–623 in your textbook, complete this graphic organizer by listing the pros and cons of building a dam.

Building a Dam	
Pros	**Cons**
•	•
•	•
•	•

Managing Natural Resources (pages 618–621)

Drawing Conclusions

As you read the lesson, think about the following question and draw a conclusion.

Which of the region's resource issues do you think should receive the most funding and attention? Explain your choice.

_____

_____

_____

_____

Terms To Know

Define or describe the following key terms from this lesson.

sustainable development

_____

_____

_____

poaching

_____

_____

Chipko

_____

_____

Academic Vocabulary

Define the following academic vocabulary words from this lesson.

assisted

_____

_____

cooperate

_____

_____

Key Points

Notes

Terms To Review

Use the following term that you studied earlier in a sentence that reflects the term's meaning from this lesson.

incentives
(Chapter 10, Section 2)

_____

_____

Seeking Solutions (page 621)

Clarifying

As you read this lesson, write down terms or concepts you find confusing. Then go back and reread the lesson to clarify the confusing parts. Write an explanation of the terms and concepts. If you are still confused, ask your teacher to clarify.

_____

_____

_____

Terms To Review

Use the following term that you studied earlier in a sentence that reflects the term's meaning from this lesson.

displaced
(Chapter 3, Section 2)

_____

_____

South Asia's Challenges (pages 622–623)

Inferring

As you read the lesson, look for clues in the descriptions and events that might help you draw a conclusion. Answer the following question.

In what ways does nuclear proliferation further complicate the already intense conflicts in South Asia?

_____

_____

_____

Terms To Know

Define or describe the following key terms from this lesson.

nuclear proliferation
_____
_____
_____

Dalits
_____
_____

Academic Vocabulary

Define the following academic vocabulary words from this lesson.

violating
_____
_____
_____

sums
_____
_____

Terms To Review

Use the following terms that you studied earlier in a sentence that reflects the term's meaning from this lesson.

ethnic
(Chapter 6, Section 3)
_____
_____

nuclear
(Chapter 11, Section 1)
_____
_____

Promise and Possibility (page 623)

Skimming

Read the title and quickly look over the lesson to get a general idea of the lesson's content. Then write a sentence or two explaining what the lesson is about.

_____
_____
_____
_____

Terms To Review

Use the following term that you studied earlier in a sentence that reflects the term's meaning from this lesson.

ensuring
(Chapter 19, Section 1)

_____

_____

Section Wrap-up

Now that you have read the section, write the answers to the questions that were included in Setting a Purpose for Reading *at the beginning of the lesson.*

How is South Asia handling the complex task of managing its rich natural resources?

_____

_____

_____

What environmental challenges does South Asia face in the years ahead?

_____

_____

_____

How do geographic factors impact the political and economic challenges of South Asia's future?

_____

_____

_____

Chapter 26, Section 1
The Land

(Pages 645–650)

Reason To Read

Setting a Purpose for Reading Think about these questions as you read:
- How are East Asia's landforms affected by the region's location on the Ring of Fire?
- How do the landforms of China differ from the rest of East Asia?
- What important natural resources are present in the region?

Main Idea

As you read pages 645–650 in your textbook, complete this graphic organizer by identifying the three tectonic plates that form the Ring of Fire.

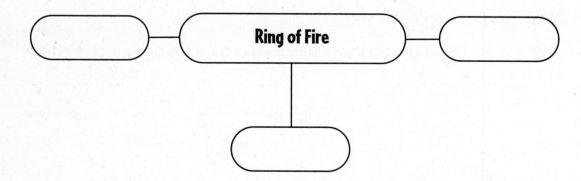

Ring of Fire

Land and Sea (pages 645–647)

Interpreting

As you read, think about the efects of the Ring of Fire on East Asia. Ask yourself, how has the Ring of Fire helped shape East Asia and what effects does it continue to have?

_____

_____

Terms To Know

Define or describe the following key terms from this lesson.

archipelago ⟩

_____

_____

tsunami ⟩

_____

_____

Places To Locate

Fill in the blank with the correct place from this lesson from the list below. One place will be used twice.

Mongolia   Macau   South China Sea

Korean Peninsula   Japan   Hong Kong

1. _____ is surrounded by the Sea of Okhotsk on the north, the Sea of Japan and the East China Sea on the west, and the Philippine Sea on the south.

2. China's northern neighbor is the country of _____ .

3. _____ is an island chain.

4. The _____ separates the Sea of Japan from the Yellow Sea.

5. _____ and _____ are two busy ports on China's southern coast.

6. The _____ stretches south from the island of Taiwan to the Philippines and the peninsula of Southeast Asia.

Terms To Review

Use the following term that you studied earlier in a sentence that reflects the term's meaning from this lesson.

occupies
(Chapter 14, Section 1)

_____

_____

Mountains, Highlands, and Lowlands *(pages 647–648)*

Previewing

Preview the lesson to get an idea of what is ahead. First, skim the lesson. Then write a sentence or two explaining what you think you will be learning. After you have finished reading, revise your statements as necessary.

_____

_____

_____

_____

Terms To Review

Use the following terms that you studied earlier in a sentence that reflects the term's meaning from this lesson.

movement
(Chapter 1, Section 1)

_____

_____

ranges
(Chapter 1, Section 2)

_____

_____

River Systems (pages 648–649)

Outlining   *Complete this outline as you read:*

I. China's Rivers

    A. _____

    B. _____

    C. _____

II. Rivers in Japan and Korea

    A. _____

    B. _____

III. The Power of Wind and Water

    A. _____

Terms To Know   *Define or describe the following key term from this lesson.*

loess >
_____
_____

Academic Vocabulary   *Define the following academic vocabulary word from this lesson.*

sites >
_____
_____

Terms To Review   *Use the following term that you studied earlier in a sentence that reflects the term's meaning from this lesson.*

delta
(Chapter 20, Section 1) >
_____
_____

Natural Resources *(pages 649–650)*

Predicting

Read the title and main headings of the lesson. Write a statement predicting what the lesson is about and what will be included in the text. As you read, adjust or change your prediction if it does not match what you learn.

_____

_____

_____

Terms To Review

Use the following terms that you studied earlier in a sentence that reflects the term's meaning from this lesson.

available
(Chapter 2, Section 3)

_____

_____

_____

distributed
(Chapter 8, Section 1)

_____

_____

Section Wrap-up

Now that you have read the section, write the answers to the questions that were included in Setting a Purpose for Reading *at the beginning of the lesson.*

How are East Asia's landforms affected by the region's location on the Ring of Fire?

_____

_____

How do the landforms of China differ from the rest of East Asia?

_____

_____

What important natural resources are present in the region?

_____

_____

Chapter 26, Section 2
Climate and Vegetation
(Pages 651-655)

Reason To Read

Setting a Purpose for Reading Think about these questions as you read:
- What accounts for East Asia's wide variety of climates?
- How do winds, ocean currents, and mountains influence the climates of East Asia?
- What conditions cause the extreme climates in much of China?
- What kinds of natural vegetation are found in East Asia's varied climate regions?

Main Idea

As you read pages 651-655 in your textbook, complete this graphic organizer by using the major headings of the section to create an outline.

I. Climate regions

    A. _____

    B. _____

    C. _____

    D. _____

    E. _____

II. Monsoons

    A. Qin Ling Mountains _____

III. Ocean Currents

    A. Japan Current _____

    B. Typhoons_____

Climate Regions *(pages 651–654)*

Visualizing

Visualize the information described is this lesson to help you understand and remember what you have read. First, read the lesson. Next, ask yourself, What would the island of Hainan look like? *Finally, write a description of the pictures you visualized on the lines below.*

_____

_____

_____

Places To Locate

Explain why the following places from this lesson are important.

Taiwan

_____

_____

Hainan

_____

_____

Terms To Review

Use the following terms that you studied earlier in a sentence that reflects the term's meaning from the lesson.

rain shadow
(Chapter 3, Section 2)

_____

_____

timberline
(Chapter 5, Section 2)

_____

_____

 Key Points

 Notes

Monsoons *(pages 654–655)*

Interpreting

As you read, think about the effects of monsoons on East Asia. Ask yourself, what economic effects would occur if the summer monsoon arrived months late in China?

_____

_____

_____

_____

Terms To Know

Define or describe the following key term from this lesson.

> **monsoons**

_____

_____

Academic Vocabulary

Define the following academic vocabulary words from this lesson.

> **approximately**

_____

_____

> **intense**

_____

_____

Terms To Review

Use the following term that you studied earlier in a sentence that reflects the term's meaning from this lesson.

> **prevailing winds**
> (Chapter 3, Section 2)

_____

_____

Notes

Ocean Currents *(page 655)*

Synthesizing

As you read, think about the main ideas of the lesson. Ask yourself, how do ocean currents affect East Asia's climate?

_____

_____

_____

_____

Terms To Know

Define or describe the following key terms from this lesson.

Japan Current >
_____

_____

typhoon >
_____

_____

Terms To Review

Use the following terms that you studied earlier in a sentence that reflects the term's meaning from this lesson.

currents
(Chapter 3, Section 2) >
_____

_____

hurricanes
(Chapter 5, Section 2) >
_____

_____

Now that you have read the section, write the answers to the questions that were included in Setting a Purpose for Reading *at the beginning of the lesson.*

What accounts for East Asia's wide variety of climates?

_____

_____

How do winds, ocean currents, and mountains influence the climates of East Asia?

_____

_____

What conditions cause the extreme climates in much of China?

_____

_____

What kinds of natural vegetation are found in East Asia's varied climate regions?

_____

_____

Chapter 27, Section 1
Population Patterns
(Pages 661–665)

Reason To Read

Setting a Purpose for Reading Think about these questions as you read:
• What ethnic groups make up East Asia's populations?
• In what country do the majority of East Asians live?
• How is the population in East Asia distributed?

Main Idea

As you read pages 661–665 in your textbook, complete this graphic organizer by describing China's "one-child" policy.

China's "One-Child" Policy
•   •   •   •

Human Characteristics *(pages 661–663)*

Responding

As you read this lesson, think about what attracts your attention as you read. Write down facts you find interesting or surprising in the lesson.

_____
_____
_____
_____
_____

Terms To Know

Define or describe the following key terms from this lesson.

aborigines
_____
_____
_____

homogeneous
_____
_____

Academic Vocabulary

Define the following academic vocabulary word from this lesson.

civil war
_____
_____

Terms To Review

Use the following terms that you studied earlier in a sentence that reflects the term's meaning from this lesson.

cultures
(Chapter 1, Section 2)
_____
_____

trace
(Chapter 9, Section 3)
_____
_____

Where East Asians Live *(page 663)*

Previewing

Preview the lesson to get an idea of what is ahead. First, skim the lesson. Then write a sentence or two explaining what you think you will be learning. After you have finished reading, revise your statements as necessary.

_____

_____

_____

Places To Locate

Write the letter of the correct location in Group B next to the correct place from this lesson in Group A. One location will not be used.

Group A

_____ **1.** Taipei

_____ **2.** Seoul

_____ **3.** Pyongyang

_____ **4.** Tokaido corridor

_____ **5.** Tokyo

Group B

a. the capital of North Korea

b. the capital of South Korea

c. the capital of Japan

d. a series of cities crowded together on the main island of Honshu

e. a region that includes a central city and its surrounding suburbs

f. the capital of Taiwan

Terms To Review

Use the following terms that you studied earlier in a sentence that reflects the term's meaning from this lesson.

urbanization
(Chapter 6, Section 1)

_____

_____

adapted
(Chapter 21, Section 3)

_____

_____

 **Key Points**

 **Notes**

Migration (page 664)

Inferring

As you read the lesson, look for clues in the descriptions of events that might help you draw a conclusion to the following question.

How might population growth and the continued migration of people from rural to urban areas affect East Asia's agricultural future?

_____

_____

_____

_____

Terms To Review

Use the following term that you studied earlier in a sentence that reflects the term's meaning from this lesson.

external
(Chapter 12, Section 1)

_____

_____

Challenges of Growth (page 665)

Predicting

Read the title and main headings of the lesson. Write a statement predicting what the lesson is about and what will be included in the text. As you read, adjust or change your prediction if it does not match what you learn.

_____

_____

_____

Terms To Review

Use the following terms that you studied earlier in a sentence that reflects the term's meaning from this lesson.

factor
(Chapter 4, Section 2)

_____

_____

statistics
(Chapter 22, Section 2)

Section Wrap-up

Now that you have read the section, write the answers to the questions that were included in **Setting a Purpose for Reading** *at the beginning of the lesson.*

What ethnic groups make up East Asia's populations?

_____

_____

In what country do the majority of East Asians live?

_____

_____

How is the population in East Asia distributed?

_____

_____

Chapter 27, Section 2
History and Government

(Pages 668–672)

Reason To Read

Setting a Purpose for Reading Think about these questions as you read:
- Where did East Asia's ideas and traditions originate?
- How did East Asia first react to contact with the West?
- What major wars and revolutions occurred in East Asia?

Main Idea

As you read pages 668–672 in your textbook, complete this graphic organizer by filling in details about the Zhou Dynasty.

The Zhou Dynasty
•
•
•
•

Ancient East Asia (pages 668–670)

Sequencing

As you read, number the following events in the correct order.

_____ Zhou dynasty rules China.

_____ Shang dynasty takes power in the North China Plain.

_____ Culture hearth emerges in China.

_____ First section of the Great Wall of China is built.

_____ Qing dynasty rules China.

_____ Naval explorer, under Ming dynasty, reaches East Africa.

Terms To Know

Fill in the blank with the correct term from this lesson from the list below. One term will be used twice.

> **culture hearth**

> **dynasty**

> **clans**

> **shogun**

> **samurai**

1. Yoritomo Minamoto became Japan's first _____, or military ruler.

2. Historical records in China were first kept under the ruling family called the Shang _____.

3. In the A.D. 400s, Japan was ruled by _____, or family groups.

4. China became East Asia's _____ , or center from which ideas and practices spread to surrounding areas.

5. In Japan the _____ were professional warriors.

6. After the Shang, the Zhou _____ was the ruling family in China.

Places To Locate

Explain why the following place from this lesson is important.

> **Great Wall of China**

_____

_____

Academic Vocabulary

Circle the letter of the word or phrase that has the closest meaning to the underlined word from this lesson.

1. Daoism is a <u>philosophy</u> of living in simplicity and harmony with nature.

   a. community   **b.** study of people   **c.** way of thinking

Terms To Review

Use the following terms that you studied earlier in a sentence that reflects the term's meaning from this lesson.

conduct
(Chapter 4, Section 4)

_____
_____

tributary
(Chapter 5, Section 1)

_____
_____

Contact With the West (page 670)

Synthesizing

As you read, think about the main ideas of the lesson. Combine the ideas in this lesson to answer the following question.

What events led to the modernization of Japan?

_____
_____
_____
_____

Places To Locate

Explain why the following place from this lesson is important.

Guangzhou

_____

Terms To Review

Use the following terms that you studied earlier in a sentence that reflects the term's meaning from this lesson.

economy
(Chapter 1, Section 2)

_____
_____
_____

rejected
(Chapter 24, Section 2)

_____
_____

Modern East Asia *(pages 670–672)*

Drawing Conclusions

As you read the lesson, look for clues to help you answer the following question:

How did Japan build an empire in the early 1900s, and how did the empire come to an end?

_____
_____

Terms To Review

Use the following terms that you studied earlier in a sentence that reflects the term's meaning from this lesson.

overseas
(Chapter 9, Section 1)

_____
_____

parallel
(Chapter 20, Section 1)

_____
_____

 Notes

 Section Wrap-up

Now that you have read the section, write the answers to the questions that were included in Setting a Purpose for Reading *at the beginning of the lesson.*

Where did East Asia's ideas and traditions originate?

_____

_____

How did East Asia first react to contact with the West?

_____

_____

What major wars and revolutions occurred in East Asia?

_____

_____

Chapter 27, Section 3
Cultures and Lifestyles
(Pages 673–679)

Reason To Read

Setting a Purpose for Reading Think about these questions as you read:
- What languages do the peoples of East Asia speak?
- What religions and philosophies do many people of East Asia follow?
- How do the standards of living of East Asians compare with one another?
- How does education in East Asia compare with education in North America?
- What traditional arts make East Asia unique?

Main Idea

As you read pages 673–679 in your textbook, complete this graphic organizer by creating an outline.

I. East Asia's Languages

    A. _____

    B. _____

II. Religion and Philosophy

III. Standard of Living

    A. _____

    B. _____

IV. Education and Health

    A. _____

    B. _____

V. Leisure Activities

    A. _____

    B. _____

VI. The Arts

    A. _____

    B. _____

    C. _____

    D. _____

East Asia's Languages *(pages 673–674)*

Scanning

Scan the lesson before you begin to read. As you glance quickly over the lines of text, look for key words or phrases that will tell you what the text will cover. Write the key words or phrases. Then use the key words and phrases to write a statement explaining the lesson content. Revise your statement when you are finished reading the lesson.

Key words or phrases >

_____

_____

_____

_____

What the lesson is about >

_____

_____

Terms To Know

Define or describe the following key term from this lesson.

ideograms >

_____

_____

Academic Vocabulary

Define the following academic vocabulary word from this lesson.

principal >

_____

_____

Religion and Philosophy *(pages 674–675)*

Determining the Main Idea

As you read, write the main idea of the lesson. Review your statement when you have finished reading and revise as needed.

_____

_____

Terms To Know

Define or describe the following key terms from this lesson.

> **shamanism**

_____
_____
_____

> **lamas**

_____
_____

Standard of Living *(pages 675–676)*

Evaluating

When you evaluate something you read, you make a judgment or form an opinion. After you read this lesson, evaluate the effectiveness of the "Great Leap Forward."

_____
_____
_____

Education and Health *(pages 676–677)*

Analyzing

As you read, think about health care in East Asia. Write a short paragraph describing the kind of medical care that East Asians practice.

_____
_____
_____
_____

Terms To Know

Define or describe the following key term from this lesson.

> **acupuncture**

_____
_____

 **Key Points**

 **Notes**

Leisure Activities *(page 677)*

Connecting

As you read, compare the leisure activities of East Asia with the leisure activities of the United States. Summarize your thoughts in a paragraph.

_____

_____

_____

_____

Academic Vocabulary

Define the following academic vocabulary word from this lesson.

phases

_____

_____

The Arts *(pages 678–679)*

Drawing Conclusions

As you read the lesson, write three details about the influence of East Asia's religions on its art forms. Then write a general statement based on these details.

Details

_____

_____

_____

Conclusion

_____

_____

Terms To Know

Write the letter of the correct definition in Group B next to the correct term from this lesson in Group A. One definition will not be used.

Group A

_____ **1.** haiku

_____ **2.** calligraphy

_____ **3.** pagoda

Group B

a. a style of architecture in traditional east Asian buildings marked by gracefully curved tile roofs in the tower style

b. fine, thin porcelain

c. the art of beautiful writing

d. a form of poetry that originally had only 3 lines and 17 syllables

Section Wrap-up

Now that you have read the section, write the answers to the questions that were included in Setting a Purpose for Reading at the beginning of the lesson.

What languages do the peoples of East Asia speak?

_____

_____

What religions and philosophies do many people of East Asia follow?

_____

_____

How do the standards of living of East Asians compare with one another?

_____

_____

How does education in East Asia compare with education in North America?

_____

_____

What traditional arts make East Asia unique?

_____

_____

Chapter 28, Section 1
Living in East Asia

(Pages 685–691)

Reason To Read

Setting a Purpose for Reading Think about these questions as you read:
- What types of governments and economies do East Asian countries have?
- What economic activities play an important role in East Asia?
- How are other countries in the region challenging Japan's economic dominance?
- How are the countries of East Asia economically interdependent?

Main Idea

As you read pages 685–691 in your textbook, complete this graphic organizer by filling in the East Asian members of the Asia-Pacific Economic Cooperation Group (APEC).

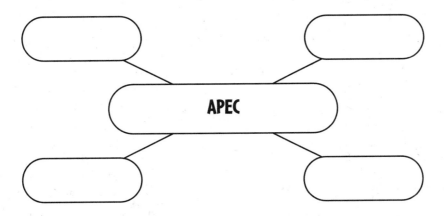

Political and Economic Systems (pages 685–686)

Reviewing

As you read the lesson, list the names of the countries of East Asia and the political and economic systems they have today.

_____

_____

_____

Terms To Know

Define or describe the following key term from this lesson.

command systems

_____

_____

Terms To Review

Use the following term that you studied earlier in a sentence that reflects the term's meaning from this lesson.

shifted
(Chapter 17, Section 1)

_____

_____

Agriculture (pages 686–687)

Analyzing

As you read the lesson, think about its organization and main ideas. Then write a sentence explaining the organization, and list the main ideas.

Organization

_____

_____

Main Ideas

_____

_____

_____

 **Key Points**

 **Notes**

Terms To Know

Define or describe the following key terms from this lesson.

communes _____
_____
_____

cooperatives _____
_____
_____

Terms To Review

Use the following term that you studied earlier in a sentence that reflects the term's meaning.

adapting
(Chapter 21, Section 3)
_____
_____

Industry *(pages 687–689)*

Responding

As you read this lesson, think about what attracts your attention as you read. Write down facts you find interesting or surprising in the lesson.

_____
_____
_____
_____

Academic Vocabulary

Circle the letter of the word or phrase that has the closest meaning to the underlined word from this lesson.

About 60 percent of Taiwan's people work in <u>finance</u> and communications.

a. farming   **b.** banking   **c.** marketing

Terms To Review

Use the following term that you studied earlier in a sentence that reflects the term's meaning from this lesson.

> **market economy**
> (Chapter 4, Section 3)

_____

_____

Trade _(pages 689–690)_

Skimming

Read the title and quickly look over the lesson to get a general idea of the lesson's content. Then write a sentence or two explaining what the lesson is about.

_____

_____

_____

Terms To Know

Write the letter of the correct definition in Group B next to the correct term from this lesson in Group A. One definition will not be used.

Group A

_____ **1.** Asia-Pacific Economic Cooperation Group (APEC)

_____ **2.** trade surplus

_____ **3.** trade deficit

_____ **4.** dissidents

_____ **5.** economic sanctions

_____ **6.** World Trade Organization (WTO)

Group B

a. imports exceed exports

b. trade restrictions

c. value of imports and exports is equal

d. a trade group whose members are Japan, China, South Korea, and Taiwan

e. citizens who speak out against government policies

f. exports exceed imports

g. an international body that oversees trade agreements and settles trade disputes among countries

Transportation and Communications (pages 690–691)

Interpreting

Think about transportation networks in East Asia. Ask yourself, why are transportation systems in Japan, South Korea, and Taiwan more developed than in the other countries in the region?

_____

_____

_____

Terms To Know

Define or describe the following key term from this lesson.

_____

merchant marine

_____

_____

Section Wrap-up

Now that you have read the section, write the answers to the questions that were included in Setting a Purpose for Reading *at the beginning of the lesson.*

What types of governments and economies do East Asian countries have?

_____

_____

What economic activities play an important role in East Asia?

_____

_____

How are other countries in the region challenging Japan's economic dominance?

_____

_____

_____

How are the countries of East Asia economically interdependent?

_____

_____

Chapter 28, Section 2
People and Their Environment

(Pages 692–697)

Reason To Read

Setting a Purpose for Reading Think about these questions as you read:
- How have industrialization and urbanization in East Asia affected the environment?
- What steps are East Asians taking to solve environmental problems?
- What naturally occurring destructive forces does East Asia regularly face?

Main Idea

As you read pages 692–697 in your textbook, complete this graphic organizer by describing the power sources of each country.

Country	Power Sources
China	
Japan	
Mongolia	
North Korea	
South Korea	
Taiwan	

The Power Dilemma (pages 692–693)

Drawing Conclusions

As you read the lesson, think about the following question and draw a conclusion.

What are some possible positive and negative aspects of using nuclear power in East Asia?

_____

_____

Academic Vocabulary

Define the following academic vocabulary word from this lesson.

devices

_____

_____

Terms To Review

Use the following term that you studied earlier in a sentence that reflects the term's meaning from this lesson.

global warming
(Chapter 3, Section 1)

_____

_____

Environmental Concerns (pages 693–695)

Clarifying

As you read this lesson, write down terms or concepts you find confusing. Then go back and reread the lesson to clarify the confusing parts. Write an explanation of the terms and concepts. If you are still confused, write your questions down and ask your teacher to clarify.

_____

_____

_____

Terms To Know

Define or describe the following key terms from this lesson.

desertification

_____

_____

_____

chlorofluoro-carbons (CFCs)

_____

_____

_____

Terms To Review

Use the following term that you studied earlier in a sentence that reflects the term's meaning from this lesson.

clear-cutting
(Chapter 7, Section 2)

_____

_____

Managing Ocean Resources *(pages 695–696)*

Inferring

As you read the lesson, look for clues in the descriptions and events that might help you draw a conclusion to the following question.

How can countries in overfished sea areas obtain seafood?

_____

_____

Terms To Know

Define or describe the following key term from this lesson.

aquaculture

_____

_____

Terms To Review

Use the following term that you studied earlier in a sentence that reflects the term's meaning from this lesson.

target
(Chapter 16, Section 2)

_____
_____

Natural Disasters *(pages 696–697)*

Summarizing

As you read, complete the following sentences. Doing so will help you summarize the lesson.

1. To control flooding along the Yellow and Yangtze Rivers, networks

of _____ and _____ have been built to

transport or redirect water quickly.

2. China's _____ is under construction upriver from Wuhan

on the Yangtze River to create a huge reservoir about 400 miles long.

3. Most East Asian countries experience destructive _____

such as the ones that struck Taiwan in late 1999.

4. Undersea volcanoes or earthquakes can cause _____ .

Academic Vocabulary

Define the following academic vocabulary word from this lesson.

crucial

_____
_____

Terms To Review

Use the following terms that you studied earlier in a sentence that reflects the term's meaning from this lesson.

tsunamis
(Chapter 26, Section 1)

_____
_____

typhoons
(Chapter 26, Section 2)

Section Wrap-up

Now that you have read the section, write the answers to the questions that were included in Setting a Purpose for Reading *at the beginning of the lesson.*

How have industrialization and urbanization in East Asia affected the environment?

_____

_____

What steps are East Asians taking to solve environmental problems?

_____

_____

What naturally occurring destructive forces does East Asia regularly face?

_____

_____

Chapter 29, Section 1
The Land

(Pages 719–724)

Reason To Read

Setting a Purpose for Reading Think about these questions as you read:
- How did tectonic plate movement, volcanic activity, and earthquakes form Southeast Asia?
- Why are the region's waterways important to its peoples?
- How do rich natural resources affect Southeast Asia's economy?

Main Idea

As you read pages 719–724 in your textbook, complete this graphic organizer to create an outline.

I. Peninsulas and Islands

    A. _____

    B. _____

II. Physical Features

    A. _____

    B. _____

    C. _____

III. Natural Resources

    A. _____

    B. _____

    C. _____

    D. _____

Peninsulas and Islands (pages 719–721)

Interpreting

As you read, think about the peninsulas and islands of Southeast Asia. Ask yourself, what geologic activities created Southeast Asia?

_____

_____

Terms To Know

Fill in the blank with the correct term from this lesson from the list below. One term will be used twice.

> **cordilleras**

1. There is a series of _____ , or groups of islands, in Southeast Asia.

2. Indonesia is an _____ , or island country of Southeast Asia.

> **archipelagos**

3. When the tectonic plates collided millions of years ago, they formed

_____ , which are parallel mountain ranges and plateaus.

Academic Vocabulary

Define the following academic vocabulary word from this lesson.

> **predominantly**

_____

_____

Terms To Review

Use the following term that you studied earlier in a sentence that reflects the term's meaning from this lesson.

> **parallel**
> (Chapter 20, Section 1)

_____

_____

Physical Features *(pages 722–723)*

Reviewing

As you read the lesson, list the main physical features of Southeast Asia and briefly describe each.

_____

_____

_____

_____

Terms To Review

Use the following terms that you studied earlier in a sentence that reflects the term's meaning from this lesson.

ecosystem
(Chapter 1, Section 1)

_____

_____

monitor
(Chapter 7, Section 2)

_____

_____

Natural Resources *(pages 723–724)*

Skimming

Read the title and quickly look over the lesson to get a general idea of the lesson's content. Then write a sentence or two explaining what the lesson is about.

_____

_____

_____

Terms To Know

Define or describe the following key terms from this lesson.

flora

_____

_____

fauna

_____

 **Key Points**

 **Notes**

Academic Vocabulary

Define the following academic vocabulary word from this lesson.

commodity ▷ _____
_____

Terms To Review

Use the following terms that you studied earlier in a sentence that reflects the term's meaning from this lesson.

varied
(Chapter 2, Section 3) ▷ _____
_____

process
(Chapter 4, Section 4) ▷ _____
_____

Section Wrap-up

Now that you have read the section, write the answers to the questions that were included in Setting a Purpose for Reading *at the beginning of the lesson.*

How did tectonic plate movement, volcanic activity, and earthquakes form Southeast Asia?

_____

_____

Why are the region's waterways important to its peoples?

_____

_____

How do rich natural resources affect Southeast Asia's economy?

_____

_____

Chapter 29, Section 2
Climate and Vegetation

(Pages 725-729)

Reason To Read

Setting a Purpose for Reading Think about these questions as you read:
- What weather pattern influences the region's climates?
- What are the region's main climate types?
- What is the main type of natural vegetation found in the region?

Main Idea

As you read pages 725-729 in your textbook, complete this graphic organizer by filling in the three types of vegetation found in the region.

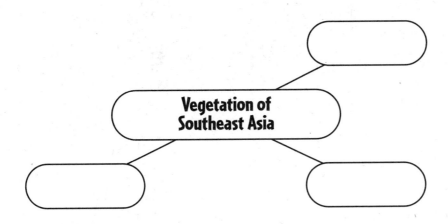

Tropical Climate Regions (pages 725–729)

Previewing

Preview the lesson to get an idea of what is ahead. First, skim the lesson. Then write a sentence or two explaining what you think you will be learning. After you have finished reading, revise your statements as necessary.

_____

_____

_____

Terms To Know

Define or describe the following key term from this lesson.

endemic

_____

_____

Places To Locate

Explain why the following place from this lesson is important.

Shan Plateau

_____

_____

Terms To Review

Use the following terms that you studied earlier in a sentence that reflects the term's meaning from the lesson.

climate
(Chapter 3, Section 1)

_____

_____

dominates
(Chapter 8, Section 2)

_____

_____

Highlands Climate (page 729)

Visualizing

Visualize the information described is this lesson to help you understand and remember what you have read. First, read the lesson. Next, ask yourself, **What would the areas with highlands climates look like?** *Finally, write a description of the pictures you visualized on the lines below.*

_____

_____

_____

_____

Places To Locate

Explain why the following places from this lesson are important.

> **Myanmar**

_____

_____

> **New Guinea**

_____

_____

> **Borneo**

_____

_____

Terms To Review

Use the following term that you studied earlier in a sentence that reflects the term's meaning from this lesson.

> **deciduous**
> (Chapter 3, Section 3)

_____

_____

Section Wrap-up

Now that you have read the section, write the answers to the questions that were included in Setting a Purpose for Reading *at the beginning of the lesson.*

What weather pattern influences the region's climates?

_____

_____

What are the region's main climate types?

_____

_____

What is the main type of natural vegetation found in the region?

_____

_____

Chapter 30, Section 1
Population Patterns

(Pages 735–739)

Reason To Read

Setting a Purpose for Reading Think about these questions as you read:
- What are the various ethnic roots of Southeast Asia's peoples?
- Why do the majority of Southeast Asians live in river valley lowlands or on coastal plains?
- How have population movements and settlement patterns affected Southeast Asia?

Main Idea

As you read pages 735–739 in your textbook, complete this graphic organizer by filling in four examples of primate cities in Southeast Asia.

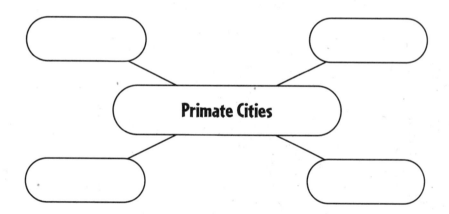

Human Characteristics (pages 735–737)

Responding

As you read this lesson, think about what attracts your attention as you read. Write down facts you find interesting or surprising in the lesson.

_____

_____

_____

_____

Terms To Review

Use the following terms that you studied earlier in a sentence that reflects the term's meaning from this lesson.

indigenous
(Chapter 9, Section 1)

_____

_____

traditions
(Chapter 4, Section 3)

_____

_____

Population Growth (page 737)

Evaluating

As you read, form an opinion about the information in this lesson. Does the quote by Jerry Adler support the following statement:

People considered to be intellectuals were often the first targets of violence by the Khmer Rouge government. Explain your answer.

_____

_____

_____

Terms To Review

Use the following terms that you studied earlier in a sentence that reflects the term's meaning from this lesson.

estimates
(Chapter 9, Section 1)

_____

_____

_____

shift
(Chapter 17, Section 1)

_____

_____

Movement to the Cities *(pages 737–738)*

Inferring

As you read the lesson, look for clues in the descriptions and events that might help you draw a conclusion. Answer the question:

What have countries in Southeast Asia done to reduce urban overcrowding?

_____

_____

Terms To Know

Define or describe the following key terms from this lesson.

urbanization

_____

_____

primate city

_____

_____

Terms To Review

Use the following terms that you studied earlier in a sentence that reflects the term's meaning from this lesson.

conflict
(Chapter 9, Section 1)

_____

_____

jobs
(Chapter 1, Section 2)

_____

_____

Outward Migrations *(pages 738–739)*

Clarifying

As you read this lesson, write down terms or concepts you find confusing. Then go back and reread the lesson to clarify the confusing parts. Write an explanation of the terms and concepts. If you are still confused, write your questions down and ask your teacher to clarify.

_____

_____

_____

_____

Terms To Review

Use the following terms that you studied earlier in a sentence that reflects the term's meaning from this lesson.

migration
(Chapter 4, Section 1)

_____

_____

contemporary
(Chapter 24, Section 3)

_____

_____

Section Wrap-up

Now that you have read the section, write the answers to the questions that were included in Setting a Purpose for Reading *at the beginning of the lesson.*

What are the various ethnic roots of Southeast Asia's peoples?

_____

_____

Why do the majority of Southeast Asians live in river valley lowlands or on coastal plains?

_____

_____

How have population movements and settlement patterns affected Southeast Asia?

_____

_____

Chapter 30, Section 2
History and Government

(Pages 740–745)

Reason To Read

Setting a Purpose for Reading Think about these questions as you read:
- How did location influence the development of empires in Southeast Asia?
- What cultural influences have affected the region's peoples?
- What events led to the independence of Southeast Asian countries?

Main Idea

As you read pages 740–745 in your textbook, complete this graphic organizer by listing the effects of Western rule in Southeast Asia.

Effects of Western Rule
•
•
•
•

Early Civilizations (pages 740–741)

Synthesizing

As you read, think about the main ideas of the lesson. Combine the ideas in this lesson to answer the following question.

What cultural traditions were followed by early civilizations in Southeast Asia?

_____

_____

_____

_____

Terms To Review

Use the following terms that you studied earlier in a sentence that reflects the term's meaning from this lesson.

domesticated
(Chapter 18, Section 2)

_____

_____

traditions
(Chapter 4, Section 3)

_____

_____

Kingdoms and Empires (pages 741–743)

Sequencing

As you read, number the following events in the correct order.

_____ The Srivijaya Empire declined.

_____ Traders from India set up trading posts.

_____ The Khmer Empire flourished.

_____ The kingdom of Funan is established.

Terms To Review

Use the following terms that you studied earlier in a sentence that reflects the term's meaning from this lesson.

routes
(Chapter 12, Section 2)

_____

_____

_____

philosophy
(Chapter 27, Section 2)

_____

_____

_____

Western Colonization (pages 743)

Drawing Conclusions

As you read the lesson, look for clues to help you answer the following question:

How did colonization by Western countries affect the region?

Conclusion

_____

_____

Terms To Know

Define or describe the following key terms from this lesson.

spheres of influence

_____

_____

buffer state

_____

_____

Academic Vocabulary

Define the following academic vocabulary word from this lesson.

neutral

_____

_____

 **Key Points**

 **Notes**

Terms To Review

Use the following terms that you studied earlier in a sentence that reflects the term's meaning from this lesson.

altered
(Chapter 2, Section 2)

_____
_____
_____

cash crops
(Chapter 22, Section 1)

_____
_____
_____

Struggle for Freedom (page 743–745)

Drawing Conclusions

As you read the lesson, look for clues to help you answer the following question:

Why did political conflicts and wars rage through Southeast Asia after independence?

Conclusion

_____
_____

Terms To Know

Define or describe the following key term from this lesson.

maritime

_____
_____

Academic Vocabulary

Define the following academic vocabulary word from this lesson.

intervened

_____
_____

 Key Points

 Notes

Use the following terms that you studied earlier in a sentence that reflects the term's meaning from this lesson.

democracy
(Chapter 4, Section 3)

_____

_____

_____

republics
(Chapter 6, Section 2)

_____

_____

Section Wrap-up

Now that you have read the section, write the answers to the questions that were included in Setting a Purpose for Reading *at the beginning of the lesson.*

How did location influence the development of empires in Southeast Asia?

_____

_____

What cultural influences have affected the region's peoples?

_____

_____

What events led to the independence of Southeast Asian countries?

_____

_____

Chapter 30, Section 3
Cultures and Lifestyles

(Pages 748–753)

Reason To Read

Setting a Purpose for Reading Think about these questions as you read:
- What makes Southeast Asia such an ethnically diverse region?
- How have outside influences affected the arts in Southeast Asia?
- How do people's lifestyles reflect Southeast Asia's diversity?

Main Idea

As you read pages 748–753 in your textbook, complete this graphic organizer by filling in the countries where Buddhism is the major religion.

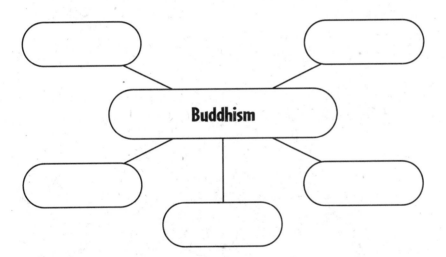

Cultural Diversity (pages 748–750)

Scanning

Scan the lesson before you begin to read. As you glance quickly over the lines of text, look for key words or phrases that will tell you what the text will cover. Write the key words or phrases. Then use the key words and phrases to write a statement explaining the lesson content. Revise your statement when you are finished reading the lesson.

Key words or phrases >

_____
_____
_____
_____

What the lesson is about >

_____
_____
_____

Terms To Review

Use the following terms that you studied earlier in a sentence that reflects the term's meaning from this lesson.

ethnic diversity
(Chapter 18, Section 1) >

_____
_____

predominant
(Chapter 29, Section 1) >

_____

The Arts (pages 750–752)

Determining the Main Idea

As you read, write the main idea of the lesson. Review your statement when you have finished reading and revise as needed.

_____
_____
_____
_____

Terms To Know

Define or describe the following key terms from this lesson.

wat >
_____
_____
_____

batik >
_____
_____
_____

Places To Locate

Explain why the following places from this lesson are important.

Irrawaddy River >
_____
_____
_____

Kuala Lumpur >
_____
_____
_____

Terms To Review

Use the following terms that you studied earlier in a sentence that reflects the term's meaning from this lesson.

mosque
(Chapter 18, Section 2) >
_____
_____
_____

adapted
(Chapter 21, Section 3) >
_____
_____
_____

Lifestyles (pages 752–753)

Synthesizing

As you read, think about the main ideas of the lesson. Ask yourself, **how has the quality of life in Southeast Asia changed since independence?**

_____
_____

Key Points

Notes

Terms To Know

Define or describe the following key term from this lesson.

> **longhouses**

_____

_____

Terms To Review

Use the following term that you studied earlier in a sentence that reflects the term's meaning from this lesson.

> **gross domestic product (GDP)**
> (Chapter 19, Section 1)

_____

_____

Section Wrap-up

Now that you have read the section, write the answers to the questions that were included in Setting a Purpose for Reading *at the beginning of the lesson.*

What makes Southeast Asia such an ethnically diverse region?

_____

_____

How have outside influences affected the arts in Southeast Asia?

_____

_____

How do people's lifestyles reflect Southeast Asia's diversity?

_____

_____

Chapter 31, Section 1
Living in Southeast Asia
(Pages 759–765)

Reason To Read

Setting a Purpose for Reading Think about these questions as you read:
- Why is rice farming the most important agricultural activity in Southeast Asia?
- Why are the countries in the region industrializing at different rates?
- How are the economies of Southeast Asia becoming more interdependent?

Main Idea

As you read pages 759–765 in your textbook, complete this graphic organizer by filling in the reasons why Singapore has Southeast Asia's most developed economy.

Singapore's Economy
•
•
•
•

Agriculture (pages 759–761)

Reviewing

As you read the lesson, complete the chart below by filling in the countries of Southeast Asia and their major cash crops. Review your chart when you have finished reading and revise as needed.

Southeast Asian Country	Major Cash Crops

Terms To Know

Define or describe the following key terms from this lesson.

paddies _____

_____

sickles _____

_____

subsistence crop _____

_____

cash crops _____

_____

Terms To Review

Use the following term that you studied earlier in a sentence that reflects the term's meaning from this lesson.

delta
(Chapter 20, Section 1) _____

_____

Forests and Mines (page 761)

 **Analyzing**

As you read this lesson, think about its organization and main ideas. Then write a sentence explaining the organization and list the main ideas.

Organization >

_____

_____

_____

Main Ideas >

_____

_____

_____

Terms To Know

Define or describe the following key term from this lesson.

lodes >

_____

_____

Academic Vocabulary

Define the following academic vocabulary word from this lesson.

compatible >

_____

_____

Terms To Review

Use the following terms that you studied earlier in a sentence that reflects the term's meaning.

surveying
(Chapter 8, Section 2) >

_____

_____

invested
(Chapter 11, Section 1) >

_____

_____

Industry (pages 761–763)

Responding

As you read this lesson, think about what attracts your attention as you read. Write down facts you find interesting or surprising in the lesson.

_____

_____

_____

Terms To Review

Use the following terms that you studied earlier in a sentence that reflects the term's meaning from this lesson.

service industry
(Chapter 10, Section 1)

_____

_____

hydroelectric power
(Chapter 8, Section 1)

_____

_____

Interdependence (pages 763)

Skimming

Read the title and quickly look over the lesson to get a general idea of the lesson's content. Then write a sentence or two explaining what the lesson covers.

_____

_____

_____

Terms To Know

Define or describe the following key terms from this lesson.

interdependent

_____

_____

Association of Southeast Asian Nations (ASEAN)

_____

_____

Transportation *(pages 764–765)*

Interpreting

Think about transportation networks in Southeast Asia. Ask yourself, **why is Singapore a major port?**

_____

_____

Terms To Know

Define or describe the following key term from this lesson.

free port ⟩

_____

_____

Communications *(page 765)*

Synthesizing

As you read, think about the main ideas of the lesson. Ask yourself, **do I understand more than the main ideas? Can I combine the ideas in this lesson to reach a new understanding?** *Use the information in the lesson to answer the following question.*

How might rural ways of life change in Southeast Asia as communications services are developed?

_____

_____

_____

_____

Key Points

Notes

Section Wrap-up

Now that you have read the section, write the answers to the questions that were included in Setting a Purpose for Reading *at the beginning of the lesson.*

Why is rice farming the most important agricultural activity in Southeast Asia?

_____

_____

_____

Why are the countries in the region industrializing at different rates?

_____

_____

_____

How are the economies of Southeast Asia becoming more interdependent?

_____

_____

_____

Chapter 31, Section 2
People and Their Environment

(Pages 766–771)

Reason To Read

Setting a Purpose for Reading Think about these questions as you read:
- What dangers are posed by volcanoes, floods, and typhoons in Southeast Asia?
- How has economic progress increased environmental pollution in the region?
- What efforts are underway to protect the environment in Southeast Asia?

Main Idea

As you read pages 766–771 in your textbook, complete this graphic to create an outline.

I. Nature's Might

    A. _____

    B. _____

II. Environmental Pollution

    A. _____

    B. _____

III. Logging, Farming, and Mining

    A. _____

    B. _____

IV. Environmental Protection

Nature's Might (pages 766–768)

Previewing

Preview the lesson to get an idea of what is ahead. First, skim the lesson. Then write a sentence or two explaining what you think you will be learning. After you have finished reading, revise your statements as necessary.

_____

_____

_____

_____

Terms To Know

Define or describe the following key terms from this lesson.

cyclone

_____

_____

typhoon

_____

_____

Places To Locate

Explain why the following places from this lesson are important.

Ring of Fire

_____

_____

Bali

_____

_____

Academic Vocabulary

Define the following academic vocabulary word from this lesson.

undergo

_____

_____

Key Points

Notes

Use the following terms that you studied earlier in a sentence that reflects the term's meaning from this lesson.

atmosphere
(Chapter 2, Section 1)

_____

_____

accompany
(Chapter 9, Section 3)

_____

_____

Environmental Pollution *(pages 768–769)*

Clarifying

As you read this lesson, write down terms or concepts you find confusing. Then go back and reread the lesson to clarify the confusing parts. Write an explanation of the terms and concepts. If you are still confused, write your questions down and ask your teacher to clarify.

_____

_____

_____

Academic Vocabulary

Use the following academic vocabulary word from this lesson in a sentence that reflects the term's meaning.

contribute

_____

_____

Logging, Farming, and Mining *(page 770)*

Inferring

As you read the lesson, look for clues in the descriptions and events that might help you draw a conclusion to the following question:

How has logging harmed the environment of Southeast Asia?

_____

_____

_____

Terms To Know

Define or describe the following key term from this lesson.

shifting cultivation >

_____

_____

Terms To Review

Use the following terms that you studied earlier in a sentence that reflects the term's meaning from this lesson.

deforestation
(Chapter 10, Section 2) >

_____

_____

_____

slash-and-burn
(Chapter 10, Section 2) >

_____

_____

Environmental Protection (pages 770–771)

Summarizing

As you read, complete the following sentences. Doing so will help you summarize the lesson.

1. To prevent further loss of rain forests, Thailand, Indonesia, the

   Philippines, and Malaysia have limited certain _____ exports.

2. Scientists predict that South Asia will lose many unique

   _____ because of logging.

3. Laos has tried to limit _____ _____ by
   resettling highlands peoples on more fertile and arable plains.

4. Bangkok, Thailand, is an example of _____

   _____ , caused by industrialization, crowded living
   and working areas, and the increased use of automobiles and other
   vehicles.

Key Points

Notes

Terms To Review

Use the following terms that you studied earlier in a sentence that reflects the term's meaning from this lesson.

exports
(Chapter 4, Section 4)

_____

_____

challenges
(Chapter 8, Section 1)

_____

_____

Section Wrap-up

Now that you have read the section, write the answers to the questions that were included in Setting a Purpose for Reading *at the beginning of the lesson.*

What dangers are posed by volcanoes, floods, and typhoons in Southeast Asia?

_____

_____

How has economic progress increased environmental pollution in the region?

_____

_____

What efforts are under way to protect the environment in Southeast Asia?

_____

_____

Chapter 32, Section 1
The Land
(Pages 793–798)

Reason To Read

Setting a Purpose for Reading Think about these questions as you read:
- How do mountains, plateaus, and lowlands differ in Australia and New Zealand?
- How have volcanoes and continental shelves formed the islands of Oceania?
- Why does the physical geography of Antarctica attract scientists?

Main Idea

As you read pages 793–798 in your textbook, complete this graphic organizer by describing the three types of islands in Oceania.

Island	Description
Low Island	
High Island	
Continental Island	

Australia: A Continent and a Country (pages 793–796)

Predicting

Read the title and main headings of the lesson. Write a statement predicting what the lesson is about and what will be included in the text. As you read, adjust or change your prediction if it does not match what you learn.

_____

_____

_____

Terms To Know

Define or describe the following key terms from this lesson.

artesian wells

_____

_____

coral

_____

_____

Terms To Review

Use the following terms that you studied earlier in a sentence that reflects the term's meaning from this lesson.

virtually
(Chapter 15, Section 2)

_____

_____

habitat
(Chapter 22, Section 2)

_____

_____

Oceania: Island Lands (pages 796–797)

Outlining

Complete this outline as you read:

I. Island Clusters

    A. _____

    B. _____

    C. _____

    D. _____

II. Island Types

    A. _____

    B. _____

    C. _____

Terms To Know

Define or describe the following key terms from this lesson.

atolls ⟩

_____

_____

lagoons ⟩

_____

_____

Terms To Review

Use the following terms that you studied earlier in a sentence that reflects the term's meaning from this lesson.

occur ⟩
(Chapter 2, Section 2)

_____

_____

category ⟩
(Chapter 21, Section 3)

_____

_____

New Zealand: A Rugged Landscape *(pages 797–798)*

Reviewing

As you read the lesson, list the local resources that help to meet New Zealand's energy needs.

_____

_____

_____

_____

_____

Terms To Review

Use the following terms that you studied earlier in a sentence that reflects the term's meaning from this lesson.

glaciers
(Chapter 2, Section 2)

_____

_____

fjords
(Chapter 11, Section 1)

_____

_____

Antarctica: A White Plateau *(page 798)*

Skimming

Read the title and quickly look over the lesson to get a general idea of the lesson's content. Then write a sentence or two explaining what the lesson is about.

_____

_____

_____

_____

Terms To Know

Define or describe the following key term from this lesson.

krill

_____

_____

Academic Vocabulary

Circle the letter of the word or phrase that has the closest meaning to the underlined word.

Scientists on Antarctica <u>investigate</u> weather patterns, measure environmental changes, and observe the sun and stars through an unpolluted atmosphere.

a. determine   **b.** study   **c.** support

Terms To Review

Use the following terms that you studied earlier in a sentence that reflects the term's meaning from this lesson.

percent
(Chapter 2, Section 1)

_____

_____

_____

research
(Chapter 1, Section 2)

_____

Section Wrap-up

Now that you have read the section, write the answers to the questions that were included in **Setting a Purpose for Reading** *at the beginning of the lesson.*

How do mountains, plateaus, and lowlands differ in Australia and New Zealand?

_____

_____

How have volcanoes and continental shelves formed the islands of Oceania?

_____

_____

Why does the physical geography of Antarctica attract scientists?

_____

_____

378

Chapter 32, Section 1

Chapter 32, Section 2
Climate and Vegetation

(Pages 799–803)

Reason To Read

Setting a Purpose for Reading Think about these questions as you read:

- How do variations in rainfall affect Australia's climate and vegetation?
- How does elevation affect climate patterns in New Zealand?
- What vegetation survives in the cold, dry Antarctic climate?

Main Idea

As you read pages 799–803 in your textbook, complete this graphic organizer by describing each region.

Region	Climate	Vegetation
Australia		
Oceania		
New Zealand		
Antartica		

 Notes

Australia *(pages 799–801)*

Previewing

Preview the lesson to get an idea of what is ahead. First, skim the lesson. Then write a sentence or two explaining what you think you will be learning. After you have finished reading, revise your statements as necessary.

_____

_____

_____

_____

Terms To Know

Define or describe the following key term from this lesson.

wattle

_____

_____

Academic Vocabulary

Define the following academic vocabulary word from this lesson.

framework

_____

_____

Terms To Review

Use the following terms that you studied earlier in a sentence that reflects the term's meaning from the lesson.

mixed forests
(Chapter 3, Section 3)

_____

_____

significant
(Chapter 18, Section 1)

_____

_____

Key Points

Notes

Oceania *(page 802)*

Skimming

Read the title and quickly look over the lesson to get a general idea of the lesson's content. Then write a sentence or two explaining what the lesson is about.

_____
_____
_____
_____
_____

Terms To Know

Define or describe the following key terms from this lesson.

doldrums >
_____
_____
_____

typhoons >
_____
_____

New Zealand *(pages 802–803)*

Visualizing

Visualize the information described is this lesson to help you understand and remember what you have read. First, read the lesson. Next, ask yourself, What would New Zealand look like? Finally, write a description of the pictures you visualized on the lines below.

_____
_____
_____
_____

Terms To Know

Define or describe the following key term from this lesson.

manuka >
_____
_____

Terms To Review

Use the following terms that you studied earlier in a sentence that reflects the term's meaning from this lesson.

blizzard
(Chapter 5, Section 2)

_____

_____

_____

isolation
(Chapter 8, Section 1)

_____

_____

Antarctica *(page 803)*

Synthesizing

As you read, think about the main ideas of the lesson. Ask yourself, What problems might researchers encounter in Antarctica, and how could these conditions be overcome?

_____

_____

_____

_____

_____

Terms To Know

Define or describe the following key terms from this lesson.

lichens

_____

_____

crevasses

_____

_____

Academic Vocabulary

Define the following academic vocabulary word from this lesson.

exhibits

_____

_____

Terms To Review

Use the following terms that you studied earlier in a sentence that reflects the term's meaning from this lesson.

approximately
(Chapter 26, Section 2)

_____

_____

brief
(Chapter 5, Section 2)

_____

_____

Section Wrap-up

Now that you have read the section, write the answers to the questions that were included in Setting a Purpose for Reading *at the beginning of the lesson.*

How do variations in rainfall affect Australia's climate and vegetation?

_____

_____

How does elevation affect climate patterns in New Zealand?

_____

_____

What vegetation survives in the cold, dry Antarctic climate?

_____

_____

Chapter 33, Section 1
Population Patterns
(Pages 811–815)

Reason To Read

Setting a Purpose for Reading Think about these questions as you read:
- What peoples settled in Australia and Oceania?
- How does the region's geography affect population density, distribution, and growth?
- What factors account for settlement in urban and rural areas?

Main Idea

As you read pages 811–815 in your textbook, complete this graphic organizer to create and outline.

I. Human Characteristics

    A. _____

    B. _____

    C. _____

    D. _____

II. Languages

III. Where People Live

    A. _____

    B. _____

    C. _____

    D. _____

Human Characteristics *(pages 811–813)*

Synthesizing

As you read, think about the main ideas of the lesson. Combine the ideas in this lesson to answer the following question.

How did the early peoples of Australia, Oceania, and New Zealand support themselves?

_____

_____

Places To Locate

Explain why the following place from this lesson is important.

Kiribati >

_____

_____

Terms To Review

Use the following terms that you studied earlier in a sentence that reflects the term's meaning from this lesson.

indigenous
(Chapter 9, Section 1) >

_____

_____

immigration
(Chapter 6, Section 1) >

_____

_____

Languages *(page 813)*

Questioning

As you read, write two questions about the main ideas presented in the text. After you have finished reading, write the answers to these questions.

Question 1 >

_____

_____

_____

_____

Question 2

_____
_____
_____
_____

Terms To Know

Define or describe the following key terms from this lesson.

Strine

_____
_____
_____

Pidgin English

_____
_____

Terms To Review

Use the following term that you studied earlier in a sentence that reflects the term's meaning from this lesson.

isolated
(Chapter 8, Section 1)

_____
_____

Where People Live *(pages 813–815)*

Inferring

As you read the lesson, look for clues in the descriptions and events that might you draw a conclusion. Answer the question: **What factors have contributed to Australia's large foreign population?**

_____
_____
_____
_____

 **Key Points**

 **Notes**

Places To Locate

Explain why the following places from this lesson are important.

Sydney

_____

_____

_____

Melbourne

_____

_____

Academic Vocabulary

Define the following academic vocabulary words from this lesson.

whereas

_____

_____

_____

highlights

_____

_____

Terms To Review

Use the following terms that you studied earlier in a sentence that reflects the term's meaning from this lesson.

population density
(Chapter 4, Section 1)

_____

_____

population distribution
(Chapter 4, Section 1)

_____

_____

 *Now that you have read the section, write the answers to the questions that were included in* Setting a Purpose for Reading *at the beginning of the lesson.*

What peoples settled in Australia and Oceania?

_____

_____

How does the region's geography affect population density, distribution, and growth?

_____

_____

What factors account for settlement in urban and rural areas?

_____

_____

Chapter 33, Section 2
History and Government

(Pages 816–821)

Reason To Read

Setting a Purpose for Reading Think about these questions as you read:
- What were the lifestyles of the region's indigenous peoples before colonization?
- How did colonial rule affect social, economic, and political structures?
- How do today's governments reflect the region's history?

Main Idea

As you read pages 816–821 in your textbook, complete this graphic organizer by filling in the hardships the Maori faced after British settlement in New Zealand.

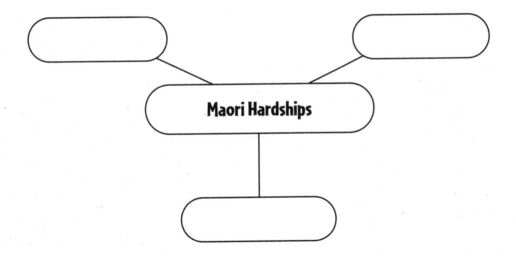

Indigenous Peoples *(pages 816–818)*

Responding

As you read this lesson, think about what attracts your attention as you read. Write down facts you find interesting or surprising in the lesson.

_____

_____

_____

_____

Terms To Know

Define or describe the following key terms from this lesson.

clans

_____

_____

_____

boomerang

_____

_____

Academic Vocabulary

Define the following academic vocabulary word from this lesson.

author

_____

_____

Terms To Review

Use the following terms that you studied earlier in a sentence that reflects the term's meaning from this lesson.

cultures
(Chapter 1, Section 2)

_____

_____

routes
(Chapter 12, Section 2)

_____

_____

European Colonization (pages 181–819)

Interpreting

As you read, think about the ways in which European settlement influenced the region. Then write a short paragraph that describes the influences.

_____

_____

_____

_____

Terms To Review

Use the following term that you studied earlier in a sentence that reflects the term's meaning from this lesson.

accurate
(Chapter 9, Section 2)

_____

_____

Struggle for Power (page 819)

Drawing Conclusions

As you read the lesson, look for clues to help you answer the following question: What events changed the course of Oceania's history?

Conclusion

_____

_____

Terms To Know

Define or describe the following key term from this lesson.

trust territories

_____

_____

Terms To Review

Use the following term that you studied earlier in a sentence that reflects the term's meaning from this lesson.

temporary
(Chapter 9, Section 1)

_____

_____

Independent Governments (pages 819–821)

Sequencing

As you read, number the following events in the correct order.

_____ The Commonwealth of Australia is formed.

_____ New Zealand becomes a self-governing dominion.

_____ Samoa becomes the first Pacific Island territory to win its freedom.

_____ Europeans first sight Antarctica.

_____ Women gain the right to vote in New Zealand.

Terms To Know

Define or describe the following key term from this lesson.

dominion

_____

_____

Academic Vocabulary

Define the following academic vocabulary words from this lesson.

prohibit

_____

_____

nonetheless

_____

_____

Places To Locate

Explain why the following places from this lesson are important.

Vanuatu

_____

_____

_____

Tonga

_____

_____

Terms To Review

Use the following terms that you studied earlier in a sentence that reflects the term's meaning from this lesson.

federal system
(Chapter 4, Section 3)

_____

_____

legally
(Chapter 9, Section 1)

_____

_____

Section Wrap-up

Now that you have read the section, write the answers to the questions that were included in **Setting a Purpose for Reading** *at the beginning of the lesson.*

What were the lifestyles of the region's indigenous peoples before colonization?

_____

_____

How did colonial rule affect social, economic, and political structures?

_____

_____

How do today's governments reflect the region's history?

_____

_____

Chapter 33, Section 3
Cultures and Lifestyles

(Pages 824–827)

Reason To Read

Setting a Purpose for Reading Think about these questions as you read:
- What role does religion play in the region's cultures?
- How have the peoples of Australia and Oceania expressed their heritages through the arts?
- How does everyday life in the region reflect cultural diversity?

Main Idea

As you read pages 824–827 in your textbook, complete this graphic organizer to create an outline.

I. A Blend of Cultures

    A. _____

    B. _____

II. Everyday Life

    A. _____

    B. _____

    C. _____

A Blend of Cultures (pages 824–826)

Evaluating

As you read, form an opinion about the information in this lesson. Does the quote by Roger Fenby support the following statement: European artists in the region began looking to the South Pacific environment for inspiration. *Explain why or why not.*

_____

_____

Academic Vocabulary

Define the following academic vocabulary words from this lesson.

visual

_____

_____

theme

_____

_____

Terms To Review

Use the following term that you studied earlier in a sentence that reflects the term's meaning from this lesson.

generation
(Chapter 21, Section 3)

_____

_____

Everyday Life (pages 826–827)

Connecting

As you read, compare the daily life in Australia, Oceania, and New Zealand. Choose a country in this region where you would most like to live. Explain, in a paragraph, why you would like to live in this country.

_____

_____

_____

_____

Terms To Know

Define or describe the following key terms from this lesson.

subsistence farming

_____

_____

_____

fale

_____

_____

Places To Locate

Explain why the following places from this lesson are important.

Papua New Guinea

_____

_____

_____

Samoa

_____

_____

Academic Vocabulary

Define the following academic vocabulary word from this lesson.

consult

_____

_____

Terms To Review

Use the following terms that you studied earlier in a sentence that reflects the term's meaning from this lesson.

bonds
(Chapter 18, Section 2)

_____

_____

service industries
(Chapter 10, Section 1)

_____

_____

 **Now that you have read the section, write the answers to the questions that were included in Setting a Purpose for Reading at the beginning of the lesson.**

What role does religion play in the region's cultures?

_____

_____

How have the peoples of Australia and Oceania expressed their heritages through the arts?

_____

_____

How does everyday life in the region reflect cultural diversity?

_____

_____

Chapter 34, Section 1
Living in Australia, Oceania, and Antarctica

(Pages 833–837)

Reason To Read

Setting a Purpose for Reading Think about these questions as you read:
- How do people in Australia, New Zealand, and Oceania make their livings?
- What role does trade play in the economies of South Pacific countries?
- What means of transportation and communications are used in the region?

Main Idea

As you read pages 833–837 in your textbook, complete this graphic organizer by filling in the developing South Pacific countries that receive much-needed income from tourism.

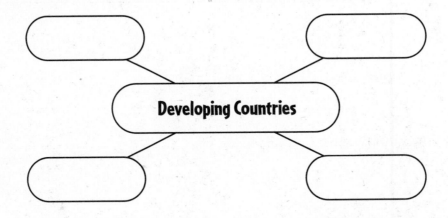

Agriculture *(pages 833–834)*

Reviewing

As you read the lesson, complete the chart below by filling in the major agricultural products of the region. Review your chart when you have finished reading and revise as needed.

Country	Major Exports
Australia	
New Zealand	
Oceania	

Terms To Know

Define or describe the following key terms from this lesson.

station ⟩ _____
_____
_____

graziers ⟩ _____
_____
_____

copra ⟩ _____
_____

Terms To Review

Use the following terms that you studied earlier in a sentence that reflects the term's meaning from this lesson.

symbol
(Chapter 9, Section 3) ⟩ _____
_____
_____

export
(Chapter 4, Section 4) ⟩ _____
_____

Mining and Manufacturing (pages 834–835)

Evaluating

When you evaluate something you read, you make a judgment or form an opinion. After you read this lesson, decide if you think Antarctica should be open to mining and explain why.

_____

_____

_____

_____

Academic Vocabulary

Define the following academic vocabulary words from this lesson.

exploited

_____

_____

core

_____

_____

Terms To Review

Use the following terms that you studied earlier in a sentence that reflects the term's meaning.

debate
(Chapter 10, Section 2)

_____

_____

enormous
(Chapter 5, Section 1)

_____

_____

Service Industries *(pages 835–836)*

Responding

As you read this lesson, think about which South Pacific country you would like to visit as a tourist. Explain why you would like to go to this country.

_____

_____

_____

_____

Terms To Review

Use the following terms that you studied earlier in a sentence that reflects the term's meaning from this lesson.

service industry
(Chapter 10, Section 1)

_____

_____

developed countries
(Chapter 4, Section 4)

_____

_____

Global Trade Links *(page 836)*

Skimming

Read the title and quickly look over the lesson to get a general idea of the lesson content. Then write a sentence or two explaining what the lesson will cover.

_____

_____

_____

_____

Terms To Review

Use the following term that you studied earlier in a sentence that reflects the term's meaning from this lesson.

supplement
(Chapter 25, Section 1)

_____

_____

Transportation and Communications (pages 836–837)

Visualizing

Visualize the information described in this lesson to help you understand and remember what you have read. First, read the lesson. Next, ask yourself, what would the South Pacific look like if I traveled around the region by airplane? Finally, write a description of the pictures you visualized on the lines below.

_____

_____

_____

Terms To Review

Use the following terms that you studied earlier in a sentence that reflects the term's meaning from this lesson.

challenge
(Chapter 8, Section 1)

_____

_____

access
(Chapter 5, Section 1)

_____

Section Wrap-up

Now that you have read the section, write the answers to the questions that were included in Setting a Purpose for Reading *at the beginning of the lesson.*

How do people in Australia, New Zealand, and Oceania make their livings?

_____

_____

What role does trade play in the economies of South Pacific countries?

_____

_____

What means of transportation and communications are used in the region?

_____

_____

Chapter 34, Section 2
People and Their Environment

(Pages 838–841)

Reason To Read

Setting a Purpose for Reading Think about these questions as you read:
- Why do Australia, Oceania, and Antarctica face many environmental challenges?
- What effects did nuclear testing have on the region?
- Why are the thinning of the ozone layer and global warming special challenges in the region?

Main Idea

As you read pages 838–841 in your textbook, complete this graphic to create an outline.

I. Managing Resources

    A. _____

    B. _____

    C. _____

II. Atmosphere and Climate

    A. Ozone Layer_____

    B. El Niño–Southern Oscillation _____

    C. Diatom_____

Managing Resources *(pages 838–841)*

Responding

As you read this lesson, think about what attracts your attention as you read. Write down facts you find interesting or surprising in the lesson.

_____

_____

_____

_____

Terms To Know

Define or describe the following key terms from this lesson.

marsupials

_____

_____

introduced species

_____

_____

food web

_____

_____

Places To Locate

Explain why the following places from this lesson are important.

Tasmania

_____

_____

Murray-Darling River Basin

_____

_____

Great Barrier Reef

_____

_____

Terms To Review

Use the following terms that you studied earlier in a sentence that reflects the term's meaning from this lesson.

coral
(Chapter 32, Section 1)

_____

_____

_____

atolls
(Chapter 32, Section 1)

_____

_____

Atmosphere and Climate *(page 841)*

Connecting

As you read the lesson, write down notes about global atmospheric and climate changes facing Australia, Oceania, and Antarctica. Use your notes to write an editorial expressing your opinions about these challenges.

_____

_____

_____

_____

_____

_____

Terms To Know

Define or describe the following key terms from this lesson.

ozone layer

_____

_____

El Niño-Southern Oscillation (ENSO)

_____

_____

diatoms

_____

_____

Key Points

Notes

Terms To Review

Use the following term that you studied earlier in a sentence that reflects the term's meaning from this lesson.

global warming
(Chapter 3, Section 1)

_____

_____

Section Wrap-up

Now that you have read the section, write the answers to the questions that were included in **Setting a Purpose for Reading** *at the beginning of the lesson.*

Why do Australia, Oceania, and Antarctica face many environmental challenges?

_____

_____

What effects did nuclear testing have on the region?

_____

_____

Why are the thinning of the ozone layer and global warming special challenges in the region?

_____

_____

CURRICULUM